IN THE WAKE OF CONTACT

Biological Responses to Conquest

IN THE WAKE OF CONTACT
Biological Responses to Conquest

Editors
CLARK SPENCER LARSEN
Department of Anthropology and
Research Laboratories of Anthropology
The University of North Carolina
Chapel Hill, North Carolina

GEORGE R. MILNER
Department of Anthropology
The Pennsylvania State University
University Park, Pennsylvania

WILEY-LISS
A JOHN WILEY & SONS, INC., PUBLICATION
New York • Chichester • Brisbane • Toronto • Singapore

Address All Inquiries to the Publisher
Wiley-Liss, Inc., 605 Third Avenue
New York, NY 10158-0012

Printed in the United States of America

Library of Congress Cataloging-in-Publication Data

In the wake of contact : biological responses to conquest / editors,
Clark Spencer Larsen, George R. Milner.
p. cm.
Includes bibliographical references (p.) and index.
ISBN 0-471-30544-8
1. Indians—Anthropometry. 2. Indians—First contact with
Europeans. 3. Paleopathology—America. 4. Epidemiology—America—
History. 5. Paleopathology—Oceania. 6. Epidemiology—Oceania—
History. I. Larsen, Clark Spencer. II. Milner, George R., 1953–
E59.A5I5 1993
614.4′2′08997—dc20 93-15716

The text of this book is printed on acid-free paper.

Contents

Contributors

Carl Armstrong
Department of Anthropology, State University of New York, College at Plattsburgh, Plattsburgh, NY 12901 **[121]**

Brenda J. Baker
Department of Sociology and Anthropology, Moorhead State University, Moorhead, MN 56563 **[35]**

Sandra V. Barnum
Department of Anthropology, University of Nebraska, Lincoln, NE 68588 **[63]**

Lynae M. Beiningen
Department of Biology, Augustana College, Sioux Falls, SD 57197 **[63]**

Mark N. Cohen
Department of Anthropology, State University of New York, College at Plattsburgh, Plattsburgh, NY 12901 **[121]**

Jerome S. Cybulski
Canadian Museum of Civilization, Hull, Quebec J8X 4H2, Canada **[75]**

Marie Danforth
Department of Sociology and Anthropology, University of Southern Mississippi, Hattiesburg, MS 39406 **[121]**

Michele T. Douglas
Department of Anthropology, University of Hawaii, Honolulu, HI 96822 **[179]**

Scott I. Fairgrieve
Department of Sociology and Anthropology, Laurentian University, Sudbury, Ontario P3E 2C6, Canada **[47]**

A. Mohammad Ghazi
Department of Geology, University of Nebraska, Lincoln, NE 68588 **[63]**

George W. Gill
Department of Anthropology, University of Wyoming, Laramie, WY 82071 **[161]**

Dale L. Hutchinson
Department of Anthropology, East Carolina University, Greenville, NC 27858 **[9]**

Keith Jacobi
Department of Anthropology, Indiana University, Bloomington, IN 47405 **[121]**

John R. Johnson
Department of Anthropology, Santa Barbara Museum of Natural History, Santa Barbara, CA 93105 **[109]**

Clark Spencer Larsen
Research Laboratories of Anthropology and Department of Anthropology, University of North Carolina, Chapel Hill, NC 27599 **[1, 21]**

Christiana E. Miewald
Department of Anthropology, University of Nebraska, Lincoln, NE 68588 **[63]**

Elizabeth Miller
Department of Anthropology, Arizona State University, Tempe, AZ 85287 **[63]**

George R. Milner
Department of Anthropology, The Pennsylvania State University, University Park, PA 16802 **[1]**

The number in brackets is the opening page number of the contributor's article.

Lynette Norr
Department of Anthropology, University of Illinois, Urbana-Champaign, IL 61820 **[9]**

Kathleen O'Connor
Department of Anthropology, University at Albany, State University of New York, Albany, NY 12222 **[121]**

Stephen D. Ousley
Department of Anthropology, University of Tennessee, Knoxville, TN 37996 **[161]**

Douglas W. Owsley
Department of Anthropology, National Museum of Natural History, Smithsonian Institution, Washington, DC 20560 **[161]**

Ann M. Palkovich
Department of Sociology and Anthropology, George Mason University, Fairfax, VA 22030 **[87]**

David M. Pendergast
Department of New World Archaeology, Royal Ontario Museum, Toronto, Ontario M5S 2C6, Canada **[135]**

Susan Pfeiffer
School of Human Biology, University of Guelph, Guelph, Ontario N1G 2W1, Canada **[47]**

Michael Pietrusewsky
Department of Anthropology, University of Hawaii, Honolulu, HI 96822 **[179]**

Karl J. Reinhard
Department of Anthropology, University of Nebraska, Lincoln, NE 68588 **[63]**

Christopher B. Ruff
Department of Cell Biology and Anatomy, School of Medicine, The Johns Hopkins University, Baltimore, MD 21205 **[21]**

Karin L. Sandness
Department of Anthropology, University of Nebraska, Lincoln, NE 68588 **[63]**

Ann L.W. Stodder
702 Pearl St., Redondo Beach, CA 90277 **[97]**

Larry Tieszen
Department of Biology, Augustana College, Sioux Falls, SD 57197 **[63]**

Douglas H. Ubelaker
Department of Anthropology, National Museum of Natural History, Smithsonian Institution, Washington, DC 20560 **[147]**

Phillip L. Walker
Department of Anthropology, University of California, Santa Barbara, CA 93106 **[109]**

Christine D. White
Department of Anthropology, University of Western Ontario, London, Ontario N6A 5C2 Canada **[135]**

Lori E. Wright
Department of Anthropology, University of Chicago, Chicago, IL 60637 **[135]**

Chapter 1

Bioanthropological Perspectives on Postcontact Transitions

Clark Spencer Larsen and George R. Milner

Research Laboratories of Anthropology and Department of Anthropology, University of North Carolina, Chapel Hill, NC 27599 (C.S.L.); Department of Anthropology, The American Museum of Natural History, New York, NY 10024 (C.S.L.); and Department of Anthropology, The Pennsylvania State University, University Park, PA 16802 (G.R.M.)

The Columbian Quincentenary prompted an outpouring of commemorative events, museum exhibits, special conferences, meeting symposia, and publications that explore various aspects of the outcome of the collision between Old and New World peoples. This response is entirely appropriate because the political, economic, social, ecological, and demographic transformations set in motion by sustained contact across the Atlantic Ocean, and soon thereafter the Pacific, collectively resulted in the world as we know it today. In fact, few academic issues have stirred the public's imagination as have controversies over the aftermath of contact between these previously isolated peoples (e.g., Thornton, 1987; Young, 1990; Fagan, 1990; Sale, 1990; Viola and Margolis, 1991; Diamond, 1992; Rouse, 1992; Black, 1992; Merbs, 1992; Daniels, 1992; Axtell, 1992a,b).

Studies of Native Americans during the first few hundred years following A.D. 1492 emphasize biological and cultural catastrophe, and the same can be said for Pacific islanders several centuries later. The significance of such research extends far beyond elucidating distant historical events. Regardless of whether studies emphasize North America or Oceania, the central issue uniting research in various disciplines is a desire to move toward a greater understanding of the biological and cultural outcome of the meeting of once-isolated populations. Of particular interest are encounters between peoples from societies that differ greatly in overall size, organizational complexity, economic underpinning, technological sophistication, ecological setting, and prior disease experience.

BIOLOGICAL ANTHROPOLOGY

Much has been written recently on early contact period changes in the population characteristics, health, and behavior of indigenous peoples. Biological anthropologists—especially bioarchaeologists—routinely deal with topics and materials that potentially can provide an important perspective on current controversies over the cultural and biological processes subsumed under labels such as First Encounters, Columbian Exchange, Columbian Conse-

In the Wake of Contact: Biological Responses to Conquest, pages 1–8

quences, or Seeds of Change (publication titles from Milanich and Milbrath [1989], Crosby [1972], Thomas [1989, 1990, 1991], and Viola and Margolis [1991]). Human skeletal remains offer special insights into the study of the human condition, a central theme of anthropology and allied disciplines. Illness, dietary composition and adequacy, habitual activities, and workload leave indelible imprints on bones and teeth. The development of means to extract such information from ancient tissues, primarily the skeleton and dentition, has added much otherwise unavailable information on human responses to changing environmental and cultural circumstances (Larsen, 1987; Iscan and Kennedy, 1989; Saunders and Katzenberg, 1992). Biological anthropologists, however, are typically underrepresented in Columbian Quincentenary publications (see Verano and Ubelaker [1992] for a notable exception to this tendency). Considering the contributions of skeletal studies to present understandings of other profound, but earlier, transformations in ways of life, it is surprising that biological anthropologists, especially osteologists, are poorly represented in the scholarly literature on the contact period. This is particularly puzzling because so many discussions revolve around biological issues (e.g., population collapse and disease).

There are several reasons for this shortcoming. First, until about 25 years ago there were relatively few trained osteologists, and they usually focused on topics other than assessments of ancient life experience. In part, this emphasis stemmed from a penchant for the characterization of cranial shape for racial identification and classification, among other such topics (Armelagos et al., 1980). Second, many of the skeletal collections dating to the early contact period, a relatively narrow window of time, have been excavated since the 1970s. Indeed, a volume like this one would not have been possible as recently as a decade ago. Third, even in regions where large samples of contact period human remains are available for study, precontact samples needed for comparative purposes are often poorly represented, thus limiting our understanding of contact period influences on life experience. Fourth, it has only been in recent years that new and more sophisticated means of data collection, analysis, and interpretation have been developed.

VOLUME CHAPTERS

This volume illustrates the variety and depth of contributions that bioarchaeologists are currently making to assessments of contact period population numbers, community composition, disease experience, cultural behavior, and the physical burdens imposed by new ways of life, especially the expropriation of native labor. Like so much of the recent contact-related literature, many chapters emphasize the devastating impact on indigenous peoples of the arrival of newcomers from the Old World. Nevertheless, the processes of accommodation to unprecedented challenges stemming from new cultural, demographic, and ecological circumstances also demand study. Many native groups displayed a remarkable resilience to pressures stemming ultimately from Old World sources, and they were able to persist, even recover, for long periods and continue as viable, albeit much changed, societies. Some such groups are examined by contributors to this volume.

Chapters representing different geographical areas and temporal periods were purposefully included in this volume (Fig. 1). The study of biological and cultural responses to contact is obviously not restricted to a specific time or place, no matter how inclusive the subject area; that is, the Americas of several hundred years ago. Two papers on Polynesia underscore the comparative potential of findings from studies related to the Columbian Quincentenary (Owsley, Pietrusewsky, and coauthors).

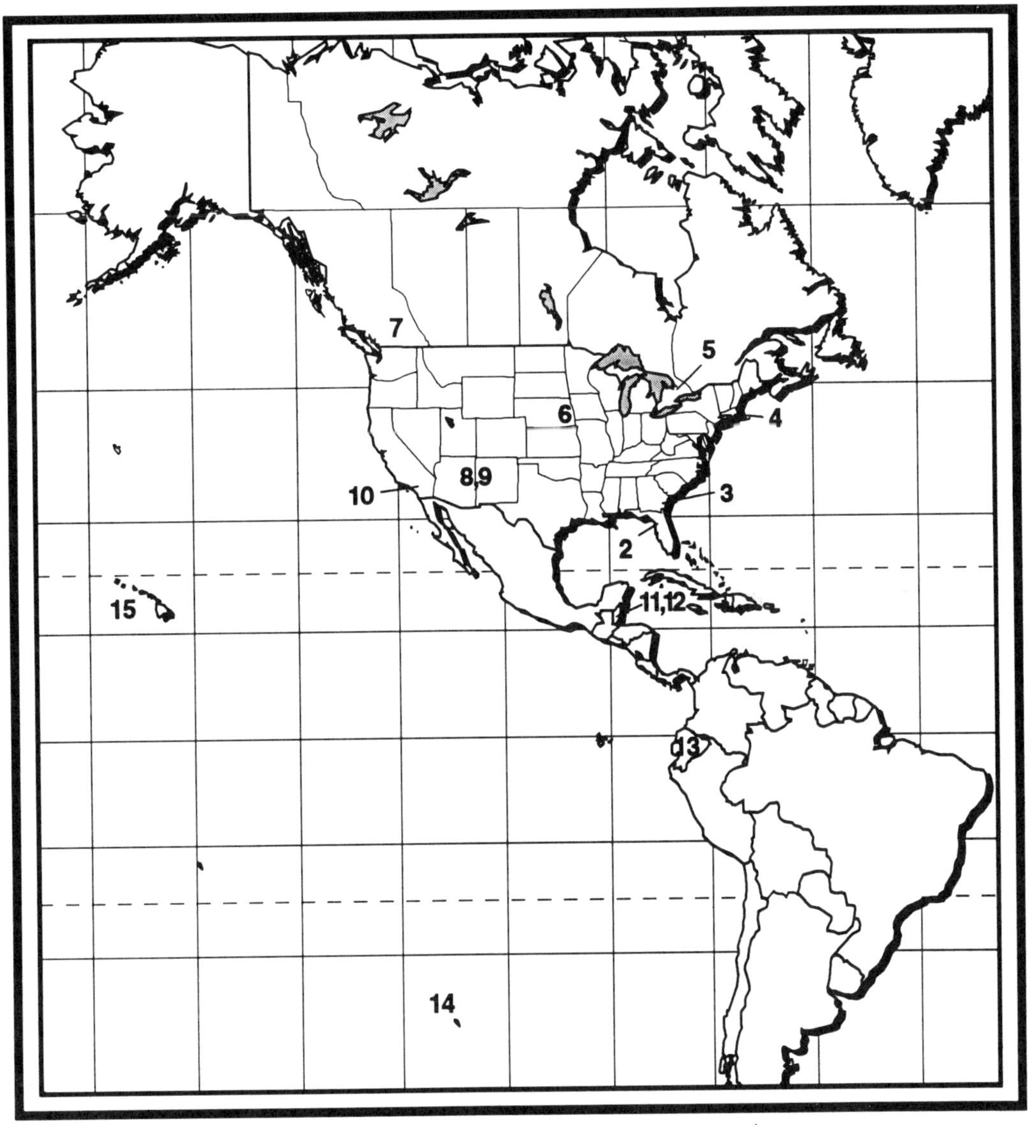

Fig. 1. Locations of skeletal collections and areas discussed in this book (indicated by chapter numbers).

Despite attempts to secure a broad geographical coverage for the volume, North America is more heavily emphasized than Central and South America. Nine chapters pertain to skeletons or historical records from the United States and Canada, whereas three report on human remains from sites in Belize and Ecuador. Uneven weighting toward North America reflects the greater amount of research dealing with archaeological human remains on this continent, including those dating to historic times. We hope the existence of this discrepancy will serve as an incentive for future workers to correct uneven geographical coverages.

This book begins with two chapters focusing on Spanish Florida, a region once known as *La Florida* (Hutchinson, Larsen, and coauthors). These chapters are followed by research on native populations in New England and northeastern Canada (Baker, Pfeiffer and Fairgrieve), the northern Plains (Reinhard and coauthors), the Northwest coast (Cybulski), and the Southwest and California (Palkovich, Stodder, Walker and Johnson). Attention then shifts southward to Belize (Cohen, White, and coauthors) and Ecuador (Ubelaker). The final two chapters highlight ongoing investigations of native populations from Polynesia, including Easter Island (Owsley and coauthors) and Hawai'i (Pietrusewsky and Douglas).

Readers will also observe that much of this book focuses on native groups that came under the influence of Spain. These chapters are based on skeletons from excavations of mission cemeteries (Larsen, White, Cohen, and coauthors) or documents from mission archives (Walker and Johnson). More skeletal samples from outlying groups clearly are needed to flesh out the altogether too spotty picture of early historic period changes in indigenous peoples who were beyond the direct reach and, hence, observation of Europeans.

The motives and aspirations of various European nations involved in exploration and conquest were quite varied, as were the ways their colonial enterprises dealt with the native populations they encountered. In northeastern North America, European contacts with native groups tended to be private and entrepreneurial with comparatively limited governmental control (Fitzhugh, 1985). In contrast, the Spanish crown and Roman Catholic church exerted a strong influence over, and closely monitored, the exploration, settlement, and exploitation of lands and peoples in the vast territories controlled by Spain in southern North America as well as in Central and South America (Thomas, 1989). In New Spain, hundreds of Catholic missions were established among native groups to encourage religious conversion and, just as important, to ensure a ready source of labor. Thus, the nature of European and Indian contact varied widely because of divergent European political and economic objectives, as well as the considerable differences that existed among the native groups themselves.

With two exceptions (Palkovich, Walker and Johnson), the chapters focus on human skeletal remains. Many recently developed osteological methods are included in this volume. Age-at-death and archaeological context provide information on sample representativeness, skeletal and dental lesions reflect disease experience, carbon and nitrogen stable isotopes clarify dietary composition, and long bone structural morphology is indicative of functional demands placed on the skeleton. Contributors also cover the nature of existing samples, the kinds of skeletal evidence available for assessments of disease experience and cultural behavior, the prospects for future studies, and the difficulties posed by that research.

Limitations stemming from the size, composition, and context of collections are all too familiar to researchers using archaeological materials, including human skeletons. Therefore, it is not surprising that many contributors direct considerable attention toward discussions of archaeological settings and sample representativeness. Bioarchaeologists will find particularly valuable the lists of skeletal collections provided in several chapters, such as those by Pietrusewsky, Pfeiffer, Ubelaker, and coauthors.

Like the volume contributors, we call attention to the potential of skeletons to contribute to knowledge about past peoples and the need for additional research to evaluate the findings reported here. Taken together, the chapters highlight at least three reasons for both caution and optimism. First, data presented in this volume are often derived from skeletons discovered only recently. Past experience indicates that skeletal collections will continue to accumulate rapidly, providing much needed large and contextually secure samples as well as more comprehensive temporal and geographical coverages. Even where the reburial of human remains is a major archaeological issue, skeletons will continue to be uncovered and studied as an outcome of construction projects and other activities affecting early mortuary sites. In fact, data summarized and questions raised in these chapters form much needed frameworks to structure future research, including work undertaken as part of salvage excavations of cemeteries. Second, judging from the past two decades we can expect that osteologists will continue to identify new kinds of data that refine old understandings and expand research horizons to encompass hitherto unaddressed questions.

Third, osteologists are continuing to wrestle with the difficult process of drawing inferences about past people from collections of skeletons. Issues of special concern include the differential preservation of bones, the size and representativeness of samples, and the recognition, quantification, and diagnosis of pathological lesions and other markers of life experience. In addition, the use of frequency data from nonsurvivors at each age (archaeological skeletons) to say something meaningful about survivors (once-living populations) is an especially problematic and largely unaddressed issue. The challenges of making sense out of skeletal data are by no means new to osteologists, although some topics have received far more attention than others (see Wood et al., 1992).

Current discussions over the kinds of bioarchaeological information that should be compiled as well as how it should be collected, handled, and interpreted are signs of a healthy, maturing discipline. Varied approaches are even apparent in the ways chosen to group skeletons for comparative purposes. That is, should skeletons from different sites be lumped together to amass large samples for statistical comparisons, or should they be kept apart because they came from separate communities with potentially different histories of disease experience (Pfeiffer and Fairgrieve, Ubelaker)? Such differences in handling skeletons parallel those in osteological studies generally, and in large part they reflect the nature of available cemetery collections and assumptions about the significance of interpopulation variability within single temporal or cultural categories. In short, at this still early stage in osteological research on the early historic period it is appropriate to assess what can, and cannot, be said about past peoples from their skeletons.

The skeletal picture of temporal changes in physical activity and disease is not easy to interpret. Roughly contemporaneous samples sometimes yield conflicting results. Occasionally, no change can be detected at the prehistoric to historic interface, despite expectations to the contrary. Furthermore, many of the most apparent alterations in activity and disease experience occurred well before contact, and historic period populations, even within single geographical areas, do not yield uniform skeletal signatures of disease experience (Cybulski, Pietrusewsky, Pfeiffer, Ubelaker, and coauthors).

These findings based on the study of archaeological skeletons underscore an obvious fact that has escaped some scholars and popular writers: namely, there are no simple dichotomies between precontact and postcontact populations in terms of skeletal or dental markers of disease experience or behavior. For example, contributors to this volume—like a great number of other researchers (e.g., Merbs, 1992; Verano and Ubelaker, 1992)—point out that the Americas were not as free of debilitating, life-threatening infectious diseases as assumed in some scholarly and popular works (e.g., Dobyns, 1983:34; Sale, 1990:160).

To be sure, skeletal data are quite different than accounts of contact-era writers; both have strengths and biases. It is hardly surprising that early chroniclers, who were all too familiar with long sea voyages and pestilential urban conditions, tended to describe the native people they encountered as being handsome, strong, and healthy (e.g., Gatschet, 1880:471). Such glowing accounts do not match the osteological picture of illness—severe and often chronic conditions were not uncommon—that has developed from numerous studies of early historic and prehistoric skeletal collections. Aside from problems stemming from the subjectivity and questionable veracity of early chroniclers, these men were relating their impressions of living people, whereas skeletal collections are obviously mortality samples.

Assessments of postcontact skeletal samples are complicated further by the presence of newly introduced, often fatal infections that either do not affect bone or produce lesions that are difficult to differentiate from the bony involvement of other conditions (Ortner and Putschar, 1985:227–229). Many premature deaths from hitherto unknown illnesses that do not involve bones should, all things being equal, reduce the number of skeletons with bone lesions from nonlethal conditions that would otherwise accumulate over individuals' lifetimes.

Several chapters illustrate how analyses of skeletal remains complement studies of historical documents (Baker, Hutchinson, Larsen, and coauthors). Information from early accounts regarding health status, behavioral patterns, and lifestyle generally can be verified through examinations of human skeletal remains, an independent source of data. The value of historical documents is highlighted by Walker and Johnson's examination of California mission records. Although all Chumash communities ultimately experienced the same outcome—drastic population decline—different missions experienced varied epidemic histories, despite regular contact among them (also see Palkovich).

Osteological findings demonstrate that the biological effects of new postcontact cultural and ecological settings were highly variable, which is not surprising considering what else is known about the native societies and European colonial endeavors of this period. The results presented in this volume represent work in progress, but it is precisely this level of specificity that is currently lacking in many influential scholarly and popular assessments of postcontact transformations in indigenous societies (e.g., Dobyns, 1983; Ramenofsky, 1987; Sale, 1990). We are not calling for a retreat to the particularistic perspectives that once characterized earlier anthropological, especially archaeological, work. Instead, we emphasize that well-controlled studies are desperately needed to resolve long-standing controversies over the magnitude, timing, rate, and causes of profound changes in cultural and demographic landscapes in the contact period. Significant gaps in the spotty temporal and geographical coverage of archaeological, including osteological, data on the immediate postcontact period are gradually being filled, as illustrated by the work of contributors to this volume.

CONCLUSION

This volume is intended as a first step in what will undoubtedly be a long process of assembling skeletal evidence on the biological ramifications of the responses of indigenous peoples to novel opportunities and challenges stemming ultimately from the Old World. Collectively, the contributors to this volume show that the biological effects of contact were not uniform, although their overall direction is clear: the sustained transoceanic interchange of humans, goods, and diseases eventually spelled disaster for the native peoples of the Americas and the Pacific islands. The work of researchers in many fields is only now beginning to put flesh on the bare bones of that altogether too general statement. What is emerging is a picture of divergent histories of response to many social and biological challenges to survival. Although such hazards were frequently new, existing threats to community existence were often exacerbated by unprecedented and sudden changes in population size and disposition, especially in the wake of deadly infectious diseases. The findings presented in this volume underscore the importance of moving beyond simplistic discussions of declining population numbers to inquiries dealing with a broad spectrum of issues, such as dietary change and nutritional deprivation, workload, warfare, and social change.

ACKNOWLEDGMENTS

This book was born out of a symposium organized for the annual meeting of the American Association of Physical Anthropologists held in Las Vegas, Nevada, in April 1992. We thank the symposium participants who, willingly and with good humor, stuck with us in the task of transforming meeting presentations into book chapters. Additionally, we thank Jerome Cybulski, Christine White, Michael Pietrusewsky, and their coauthors for contributing manuscripts on short notice following the symposium. While on the faculty at Purdue University, Larsen's fellowship from the University's Center for Social and Behavioral Science facilitated the production of this book. Sabbatical leaves from Purdue University and The Pennsylvania State University provided the time we both needed to complete the project.

We gratefully acknowledge our editor at Wiley-Liss, William F. Curtis, for his encouragement and for leading us through the many details that go into the preparation of manuscripts for publication. His help, and that of his capable staff, are much appreciated.

REFERENCES CITED

Armelagos GJ, Carlson DS, and Van Gerven DP (1980) The theoretical foundations and development of skeletal biology. In F Spencer (ed.): A History of American Physical Anthropology: 1930–1980. New York: Academic Press, pp. 305–328.

Axtell J (1992a) Beyond 1492: Encounters in Colonial North America. New York: Oxford University Press.

Axtell J (1992b) Columbian encounters: Beyond 1492. William and Mary Quarterly 44:335–360.

Black FL (1992) Why did they die? Science *258*: 1739–1740.

Crosby AW (1972) The Columbian Exchange: Biological and Cultural Consequences of 1492. Westport: Greenwood Press.

Daniels JD (1992) The Indian population of North America. William and Mary Quarterly 44:298–320.

Diamond J (1992) The arrow of disease. Discover *13*(10):6473.

Dobyns HF (1983) Their Number Become Thinned: Native American Population Dynamics in Eastern North America. Knoxville: University of Tennessee Press.

Fagan B (1990) A clash of cultures. Archaeology *43*(1):32–37.

Fitzhugh WW (1985) Commentary on part IV. In WW Fitzhugh (ed.): Cultures in Contact: The European Impact on Native Cultural Institutions in Eastern North America, A.D. 1000–1800. Washington, DC: Smithsonian Institution Press.

Gatschet AS (1880) The Timucua language. Proceedings of the American Philosophical Society *18*:465–502.

Iscan MY, and Kennedy KAR (eds.) (1989) Reconstruction of Life from the Skeleton. New York: Alan R. Liss.

Larsen CS (1987) Bioarchaeological interpretations of subsistence economy and behavior from human skeletal remains. In MB Schiffer (ed.): Advances in Archaeological Method and Theory, vol. 10. San Diego: Academic Press, pp. 339–445.

Merbs CF (1992) A new world of infectious disease. Yearbook of Physical Anthropology 35:3–42.

Milanich JT, and Milbrath S (1989) First Encounters: Spanish Explorations in the Caribbean and the United States, 1492–1570. Gainesville: University Presses of Florida.

Ortner DJ, and Putschar WGJ (1985) Identification of Pathological Conditions in Human Skeletal

Remains. Washington, DC: Smithsonian Institution Press.

Ramenofsky AF (1987) Vectors of Death: The Archaeology of European Contact. Albuquerque: University of New Mexico Press.

Rouse I (1992) The Tainos: Rise and Decline of the People Who Greeted Columbus. New Haven: Yale University Press.

Sale K (1990) The Conquest of Paradise: Christopher Columbus and the Columbian Legacy. New York: Penguin Books.

Saunders SR, and Katzenberg MA (eds.) (1992) Skeletal Biology of Past Peoples: Research Methods. New York: Wiley-Liss.

Thomas DH (ed.) (1989) Columbian Consequences, vol. 1: Archaeological and Historical Perspectives on the Spanish Borderlands West. Washington, DC: Smithsonian Institution Press.

Thomas DH (ed.) (1990) Columbian Consequences, vol. 2: Archaeological and Historical Perspectives on the Spanish Borderlands East. Washington, DC: Smithsonian Institution Press.

Thomas DH (ed.) (1991) Columbian Consequences, vol. 3: The Spanish Borderlands in Pan-American Perspective. Washington, DC: Smithsonian Institution Press.

Thornton R (1987) American Indian Holocaust and Survival: A Population History since 1492. Norman: University of Oklahoma Press.

Verano JW, and Ubelaker DH (eds.) (1992) Disease and Demography in the Americas. Washington, DC: Smithsonian Institution Press.

Viola HJ, and Margolis C (1991) Seeds of Change: A Quincentennial Commemoration. Washington, DC: Smithsonian Institution Press.

Wood JW, Milner GR, Harpending HC, and Weiss KM (1992) The osteological paradox: Problems of inferring prehistoric health from skeletal samples. Current Anthropology *33*:343–370.

Young PA (1990) Hello, Columbus! Archaeology *43*(1):2.

Chapter 2

Late Prehistoric and Early Historic Diet in Gulf Coast Florida

Dale L. Hutchinson and Lynette Norr

Department of Anthropology, East Carolina University, Greenville, NC 27858 (D.L.H.); and Department of Anthropology, University of Illinois, Urbana-Champaign, IL 61820 (L.N.)

In the southeastern United States, Europeans reported that missionized Indians were encouraged, if not forced, to focus their attention on maize agriculture, continuing a subsistence trend that began in some areas as early as A.D. 800 during the Mississippian period (Sturtevant, 1962; Smith, 1978, 1990). Spanish reliance on this native resource is well documented in the early narratives relevant to the Spanish colonies (Oviedo, 1944; Smith, 1968; Bennett, 1975). The narratives of the early exploratory *entradas* indicate, however, that maize was not utilized to the same degree by all indigenous populations initially encountered in Florida. Near the St. Johns River, for instance, the Frenchman Jacques Le Moyne reported receiving maize from the local inhabitants between A.D. 1561 and 1565 (Lorant, 1965). A similar reliance on maize prior to and during the time of initial European contact in the Apalachee region of northwestern Florida is indicated by both archaeological and ethnohistoric evidence (Milanich and Fairbanks, 1980; Hann, 1988; Scarry, 1990). Such dietary patterns contrast with those farther south in Florida.

The southern third of the Florida peninsula was occupied by Indians who paid tribute to Carlos, who it appears exerted his authority from a principal village on the southwestern coast of the state. Sixteenth-century narratives do not indicate that maize agriculture was practiced in south Florida, and most workers argue that the Calusa were entirely involved in a marine economy (Goggin and Sturtevant, 1964; Milanich et al., 1984; Marquardt, 1986, 1992; Widmer, 1988; Hann, 1991; Newsom, 1991a; Scarry and Newsom, 1992; Walker, 1992).

Human skeletons from the central Gulf coast Florida population interred at Tatham Mound provide an unusual opportunity to evaluate prehistoric and historic maize consumption in what is today Florida. The Tatham Mound remains are particularly important because the site is located in the central Gulf coast region between areas to the north and south where populations practiced radically different subsistence routines. In order to address the hypothesis that the ethnohistoric narratives are correct in reporting that maize was not of major importance in the subsistence plan of central Gulf coast populations, we conducted an analysis of the human skeletal remains for the determination of stable carbon and

In the Wake of Contact: Biological Responses to Conquest, pages 9–20

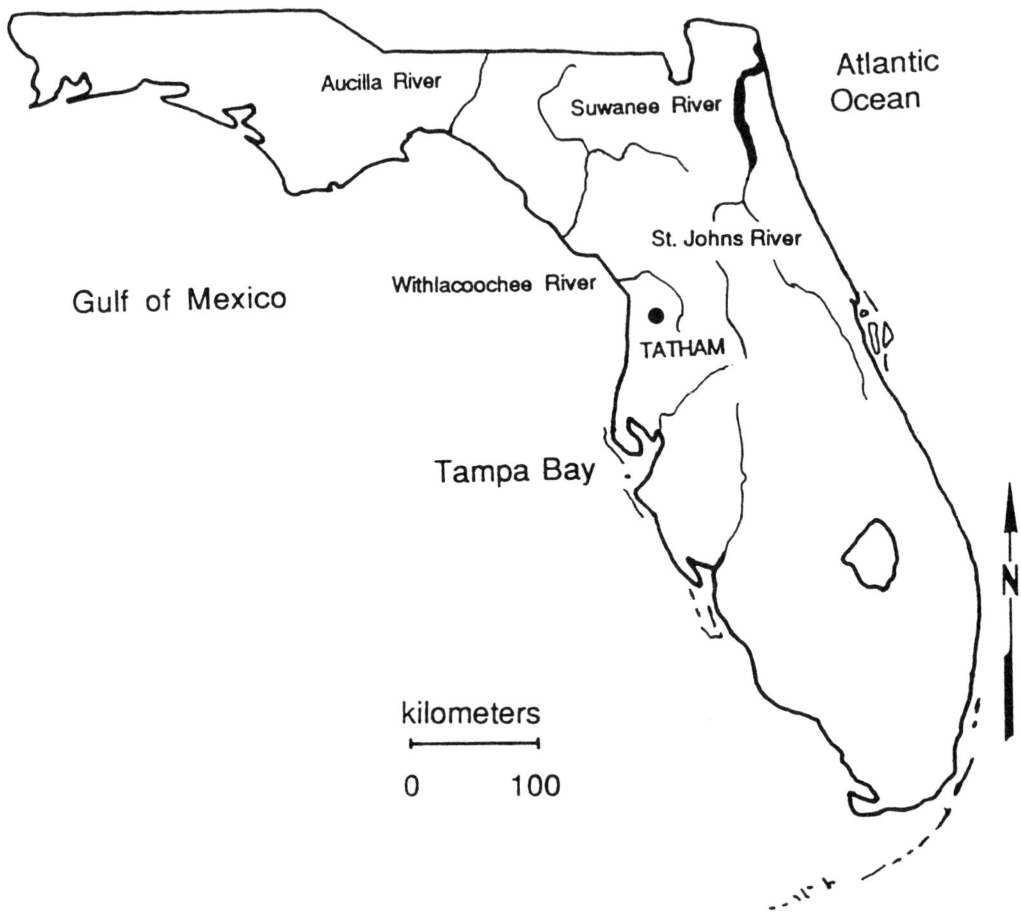

Fig. 1. Map of Florida showing major rivers and the location of Tatham Mound.

nitrogen isotope ratios in the recovered organic fraction of bone, referred to here as bone "collagen."

ARCHAEOLOGY AND ETHNOHISTORY

There is little archaeological information on the subsistence practices of populations occupying the central Gulf coast of Florida during the period from A.D. 900 to 1725, such as those in the vicinity of Tatham Mound (Fig. 1). Despite a long history of excavation (Stirling, 1930; Willey, 1949), subsistence information from the investigation of animal remains is limited to only a few sites and is often biased toward larger faunal species (Kozuch, 1986; Fitzgerald, 1987). In addition, little archaeobotanical information is available for late prehistoric Safety Harbor period sites; however, the regional database of plant remains has increased during the last decade (Newsom, 1987; Newsom and Quitmeyer, 1992).

Historic accounts from two early Spanish explorations, those of Pánfilo Narváez in 1528 and Hernando de Soto in 1539, provide some details about aboriginal life in central and northern Gulf coast Florida (Bourne, 1922; Oviedo, 1944; Smith, 1966). In particular, it appears that the large fields of maize associated with classic Mississippian settlements were absent in the central

Gulf coast area. In fact, the narratives of the de Soto expedition report little or no maize near Tampa Bay in 1539. Several days travel north of the bay, Ranjel noted that:

> . . . they came to the plain of Guaçoco, and the soldiers went into the corn fields and gathered the green corn with which they cheered themselves a little, for it was the first they had seen in that country. (Bourne, 1922:64)

Similar reports of sporadic maize fields were presented until they reached the Apalachee province to the north (Oviedo, 1944; Smith, 1968).

It was not until the Spaniards were north of what is probably the Withlacoochee River, that they found more maize:

> Next the Governor sent back some of the horsemen with mules, that had been brought from Cuba, loaded with corn and other provisions for those that were behind, since he had come upon an abundance. . . . The camp was now at Ocale, a village in a good region for corn. (Bourne, 1922:68)

Finally, when the Spanish arrived in the province of Apalachee, which was where they spent their first winter, they found abundant food:

> The Province of Apalache is very fertile and abundantly provided with supplies with much corn, kidney beans, pumpkins, various fruits, much venison, many varieties of birds, and excellent fishing near the sea; and it is a pleasant country, though there are swamps, but these have a hard sandy bottom. (Bourne, 1922:82)

It appears that maize became more abundant as the expedition moved north, and that the Spanish *entradas* first passed though an area where either the practical realities prohibited horticulture or natural resource productivity permitted a different economy. The small quantity of stored foods, such as maize, was probably an influential factor in the rapid movement of both the de Soto and Narváez expeditions through the central part of Florida. Archaeological evidence of cultural dispersion supports the view that the geographic extent of maize and the area of Safety Harbor material culture are the same. Mitchem (1989a) and others argue convincingly that the Withlacoochee River formed a natural boundary for the Safety Harbor groups.

Even in northeastern Florida, there are unresolved questions regarding the dietary significance of maize. Le Moyne, one of the French colonists who accompanied Jean Ribaut in 1564, reported large storehouses near the mouth of the St. Johns River, although he does not specify the storage of maize (Lorant, 1965). Laudonnière, another of the French colonists who accompanied Jean Ribault in 1562, reported:

> The Indians sow their corn twice a year, in March and June, replanting the same soil. This corn, from the time it is harvested is in the ground about three months, and the land is allowed to rest during the other six months of the year. (Bennett, 1975:15)

Nevertheless, archaeological evidence does not support a precontact date for maize along the St. Johns River. Botanical remains from the Hontoon site, for instance, do not include maize in the prehistoric component even though the preservation of botanical remains was very good, because they were excavated from fully water-saturated deposits (Newsom, 1987). Newsom speculates that one possibility for the sudden appearance of kernels and cobs in strata dating to the middle to late seventeenth century is an increase in maize production due to the establishment of two Franciscan missions nearby (Newsom, 1987). In fact, Newsom (1991b) reports that 16 systematic archaeobotanical studies in peninsular Florida and southeast Georgia have only found maize in contact period and later deposits.

To the south of Tampa Bay, Spanish accounts indicate that other sources of carbohydrates were utilized. Near Lake Okeechobee, in south-central Florida, Fontaneda (1868) reported that bread consumed in the region was made from locally grown roots. Although many researchers believe that the root was *Zamia*, a cycad with a starchy root/stem, Widmer (1988) presents several reasons to doubt whether *Zamia* was used prior to the arrival of the Seminoles. To complicate subsistence interpretations, maize pollen was identified at the Fort Center site, on the west side of the lake (however, see Milanich, 1987; Newsom, 1991a). The most secure contexts for that pollen are dated to A.D. 300 (Sears 1982; Sears and Sears, 1976), but no other evidence for maize has been unearthed in the region. Although Keegan (1987) attempted to resolve the issue of maize at Fort Center utilizing stable carbon and nitrogen analysis of human skeletal remains, his results indicated poor sample preservation and contamination from soil carbon, making them unusable for paleodietary reconstruction. Therefore, it seems likely that reports of early maize from Fort Center are premature, and further examination of subsistence data is required (Scarry, 1982; Griffin, 1988; Johnson, 1990).

In summary, the ethnohistoric and archaeological data indicate that Florida cannot be treated as a homogeneous natural or cultural area, and varied economic strategies likely characterized prehistoric and early historic societies. Consequently, additional information on the subsistence patterns of the people who inhabited the cove of the Withlacoochee prior to and during the first European advances into the interior is needed to evaluate the dietary significance of maize.

THE TATHAM SKELETONS

The contact period skeletons selected for analysis are from the Tatham Mound site (Mitchem, 1989b; Hutchinson, 1991). The site is located in a series of hammocks surrounded by small lakes known collectively as Lake Tsala Apopka. Bordering the site is the Withlacoochee River, a major waterway that forms a cultural boundary for the northern extent of Safety Harbor material culture. The native population associated with Tatham Mound was undoubtedly contacted directly by the Spanish, as indicated by the presence of hundreds of European artifacts at the site that date the contact period component to between A.D. 1525 and 1550 (Mitchem, 1989b). Additionally, the population interred at Tatham apparently had been subjected to some violence as indicated by skeletal elements cut by metal weapons (Fig. 2; Hutchinson, 1991).

Fig. 2. Tatham Mound skeletal elements cut by metal weapons. Top, right scapula; bottom, left humerus. Scale refers to scapula. (Photographs reproduced with permission of University Press of Florida.)

There was also a precontact component of the mound that was separated from the contact component by a sterile stratum. Artifacts associated with some of the precontact mound burials were different than those located above them, such as a copper plate with repoussé decoration, plume, and ear spool, all of which are characteristic of the Mississippian period. The artifacts indicate a date between A.D. 1000 and 1500; a radiocarbon date on a wooden post yielded an earlier date of A.D. 775–940 (Mitchem, 1989b). Carbon and nitrogen isotopic data are available for 18 contact period individuals and two precontact burials. The discussion and results of that analysis are presented below.

STABLE CARBON AND NITROGEN ISOTOPES

Certain categories of foods including plants with C_3, C_4, or CAM photosynthetic pathways (Smith and Epstein, 1971; Smith, 1972), animals feeding on a single category of those plants (Vogel, 1978; DeNiro and Epstein, 1978; van der Merwe, 1982; Ambrose and DeNiro, 1986), and animals in the terrestrial versus marine ecosystems (Chisholm et al., 1982; Schoeninger and DeNiro, 1984; Keegan and DeNiro, 1988; Norr, 1991) typically differ predictably in their stable carbon isotope ($^{13}C/^{12}C$) compositions. Plants with a C_3 photosynthetic pathway include most temperate grasses, trees, fruits, and tubers, and C_4 plants are usually tropical grasses like corn, some amaranths, chenopods, and setarias. CAM plants include many succulents and epiphytes. Seagrasses are isotopically similar to C_4 plants (Benedict et al., 1980), and algae and plankton have $^{13}C/^{12}C$ ratios like C_3 plants (Deines, 1980). Legumes versus other terrestrial plants (DeNiro and Epstein, 1981; Norr, 1991) and marine versus terrestrial fauna (Schoeninger and DeNiro, 1984; Norr, 1991) usually differ in their stable nitrogen isotope ($^{15}N/^{14}N$) composition, but reef fish may be an exception (Keegan and DeNiro, 1988). The isotopic composition of the diet is incorporated into various body tissues such as bone collagen (DeNiro and Epstein, 1978, 1981; van der Merwe, 1982; Ambrose and Norr, n.d.), and diet-related isotopic signatures may be preserved for hundreds or thousands of years in the organic fraction of bone (van der Merwe and Vogel, 1978; DeNiro, 1985; Ambrose, 1990, 1993).

Ratios of stable carbon ($^{13}C/^{12}C$) and nitrogen ($^{15}N/^{14}N$) isotopes are reported as a delta (δ) value in parts per mil (‰) relative to reference standards, the Peedee belemnite (PDB) for carbon and atmospheric nitrogen (AIR) for nitrogen. The isotopic composition of some terrestrial food resources from tropical Central America and central Gulf coast Florida are presented in Figure 3. Nitrogen isotope variation along the y axis distinguishes between most terrestrial and marine organisms, and the x axis (carbon isotopes) distinguishes terrestrial C_3, marine, and terrestrial C_4 resources.

Bone samples from the Tatham population came from the long bone shafts and ribs of adults. Clean, dry, crushed bone samples were demineralized using 0.2 molar HCl and soaked in a 0.125 molar NaOH solution prior to solubilization of collagen in 10^{-3}HCl at 95°C (DeNiro and Epstein, 1978; Ambrose, 1990; Norr, 1991). Freeze-dried collagen was placed in quartz tubing with copper oxide, copper, and silver for sealed-tube in vacuo combustion at 900°C. The resulting CO_2 and N_2 gases were cryogenically distilled on a glass vacuum line and collected in sealed Pyrex tubes. Carbon and nitrogen isotope ratios were determined separately by light-element isotope-ratio mass spectrometry. Collagen yields from clean, dry bone averaged 5%, and were as high as 20% for one individual. C:N ratios of collagen from manometric measurements of CO_2 and N_2 ranged from 3.0 to 3.4, similar to that of modern collagen (DeNiro, 1985; Ambrose, 1990).

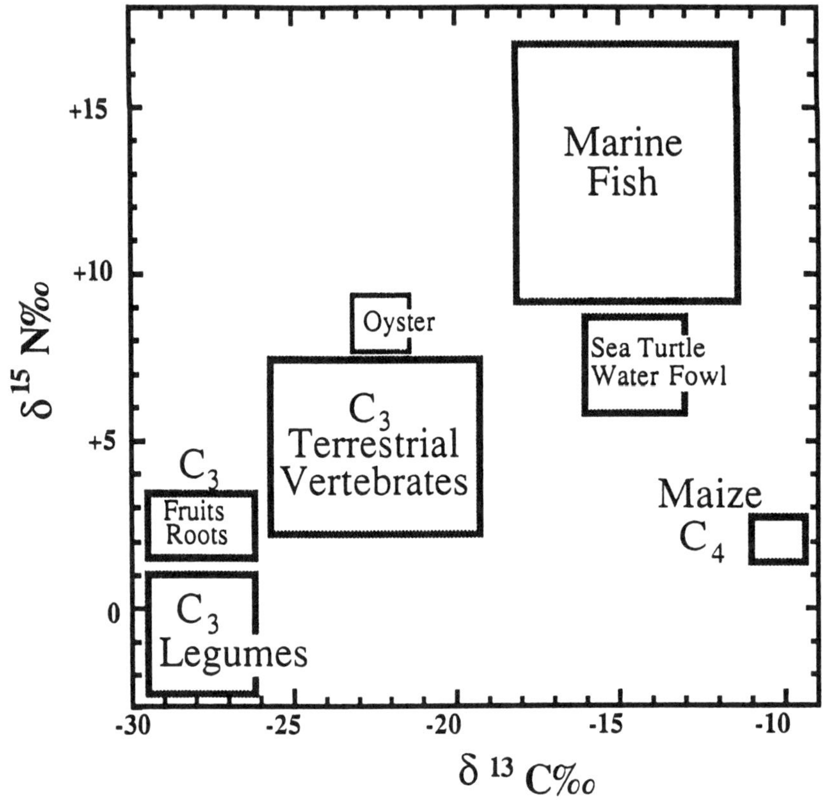

Fig. 3. Carbon and nitrogen isotopic composition of food categories (modified from Norr, n.d.).

ISOTOPIC SIGNATURES

Carbon and nitrogen isotopic results from Tatham Mound are presented in Table 1. Previous stable isotope analyses on human populations provide measures that can be used to interpret Tatham results. Schoeninger and colleagues (1983) demonstrated that isotopically different human diets can be distinguished by the analysis of human bone collagen (Fig. 4). Most distinct are the $\delta^{13}C$ and $\delta^{15}N$ values of bone collagen from individuals who consumed primarily any of the following: marine foods, C_3 plants and their animal consumers, or C_4 plants (maize). Using the results of Schoeninger et al. (1983) as reference points puts into perspective the isotopic results from mixed diets such as those of the Tatham Mound people (Figs. 5–7).

Studies have shown that neotropical coastal environments offer human populations a wide range of subsistence opportunities (Stark and Voorhies, 1978). The results obtained for the Tatham population demonstrate that C_4 pathway plants contributed little to the diet (Fig. 5). The sole outlier, an intrusive skeleton, falls closer to C_4 consuming populations, which is consistent with a later Spanish focus on maize agriculture. The two precontact period Tatham individuals had an isotopically similar diet to that of the contact period, suggesting that little change occurred from late prehistoric times to about A.D. 1550.

TABLE 1. Human Bone Collagen Stable Isotope Results from Tatham Mound, Florida

Period	Burial	Sex	%C/wt in collagen	%N/wt in collagen	C:N ratio	$\delta^{13}C$‰	$\delta^{15}N$‰
Precontact	113	Female	35.4	12.8	3.2	−17.5	+10.9
	127	Male	35.9	13.1	3.2	−20.3	+11.9
		Mean	35.6	12.9	3.2	−18.9	+11.4
		SD	0.4	0.2	0.0	2.0	0.7
Contact	1	Female	33.4	11.9	3.3	−18.8	11.8
	2, #1	Female	44.3	17.2	3.0	−13.8	10.1
	9	Male	21.8	7.4	3.4	−18.7	12.0
	14	Female	31.2	10.9	3.3	−15.6	10.9
	16	Male	31.0	16.4	3.2	−13.9	11.4
	17	Male	29.0	10.2	3.3	−15.8	11.5
	24	Unknown	27.0	9.7	3.3	−18.4	12.0
	30	Male	26.6	9.0	3.4	−17.7	11.7
	33, #1	Female	35.6	13.7	3.0	−16.9	10.8
	42, #1	Unknown	27.1	9.4	3.4	−18.1	12.0
	48, #2	Unknown	28.7	10.1	3.3	−18.6	12.0
	51	Male	31.5	11.5	3.2	−18.5	12.6
	55	Unknown	36.9	12.3	3.3	−19.1	11.9
	56, #2	Unknown	13.3	4.6	3.4	−19.7	11.7
	66	Female	25.0	8.7	3.4	−17.8	10.3
	77	Female	41.7	15.2	3.2	−15.0	10.4
	86	Female	44.4	16.1	3.2	−17.3	11.7
	120	Unknown	35.7	13.0	3.2	−18.6	11.6
		Total mean	31.3	11.5	3.3	−17.3	+11.5
		SD	7.8	3.3	0.1	1.8	0.7
		Male mean				−16.9	+11.8
		SD				2.0	0.5
		Female mean				−16.5	+10.9
		SD				1.7	0.7

In Figure 6, the Tatham results are compared to two different prehistoric marine-based populations, one C_3 marine and one C_4 marine. The C_4 marine example is from the Santa Barbara coast of California (Walker and DeNiro, 1986), and it illustrates a continuum of isotopic values ranging from terrestrial diets without maize to marine diets without maize. The C_4 marine example is from northwestern Costa Rica, and it illustrates diets consisting primarily of maize and marine fish and invertebrates (Norr, 1991). The distributions of isotopic values of bone collagen of individuals from Costa Rica and California are quite different from those of Tatham. Although the $\delta^{15}N$ values of the Florida populations are elevated above those of a terrestrial diet, it is possible, at least for Tatham, that the $\delta^{15}N$ is of freshwater origin. Schwarcz and colleagues (1985) found a similar isotopic distribution among lakeshore-dwelling prehistoric populations of southern Ontario. High $\delta^{15}N$ values result from the consumption of lake fish, and high (less negative) $\delta^{13}C$ values after A.D. 1000 reflect maize consumption (Fig. 7).

We believe the subtropical aquatic and marine environment of the Florida peninsula to be isotopically complex, and we must stress that with only limited data on the isotopic composition of local fauna and

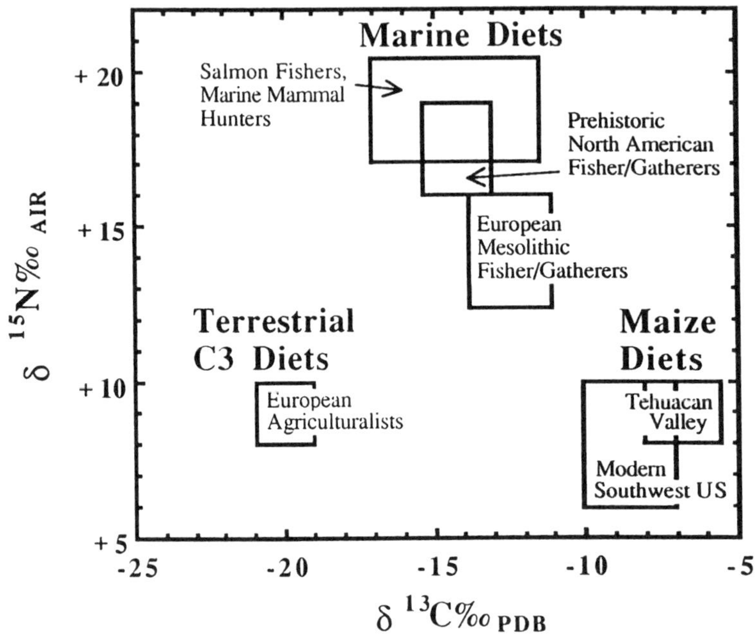

Fig. 4. Isotopic composition of bone collagen from humans with isotopically distinct diets (constructed from data in Schoeninger et al., 1983).

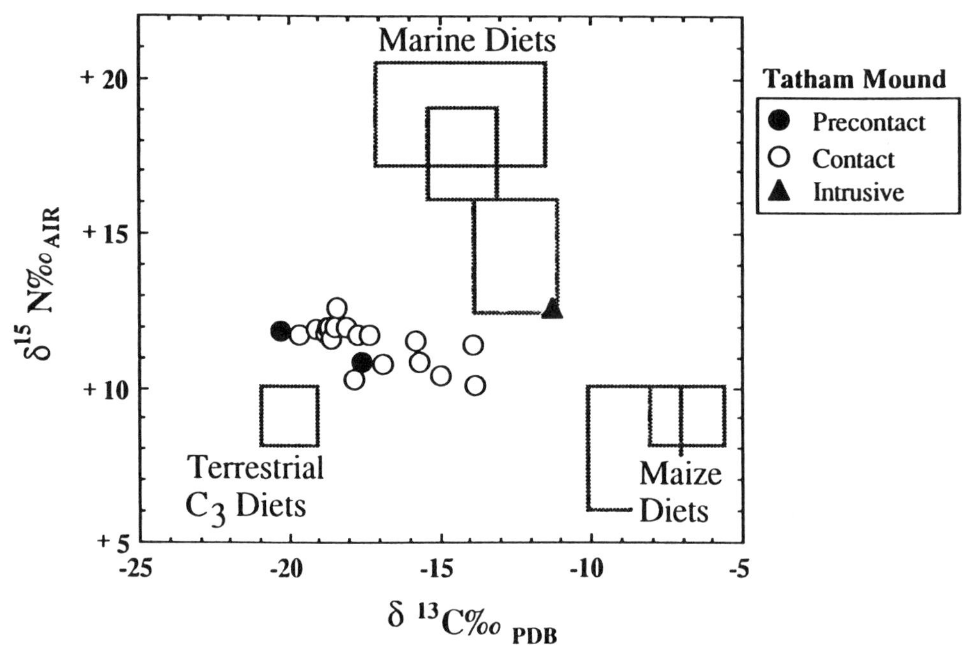

Fig. 5. Isotopic composition of human bone collagen from Tatham Mound, Florida (boxes represent data from Schoeninger et al.,1983).

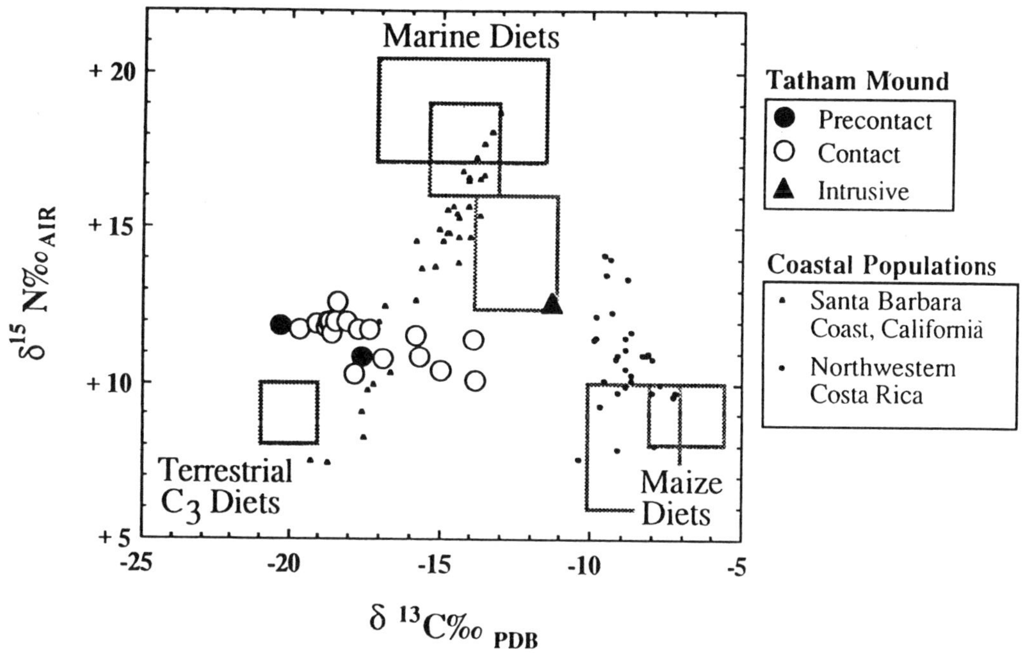

Fig. 6. Tatham Mound, Florida, compared to two coastal populations (boxes represent data from Schoeninger et al., 1983; Santa Barbara data from Walker and DeNiro, 1986; Costa Rica data from Norr, 1991).

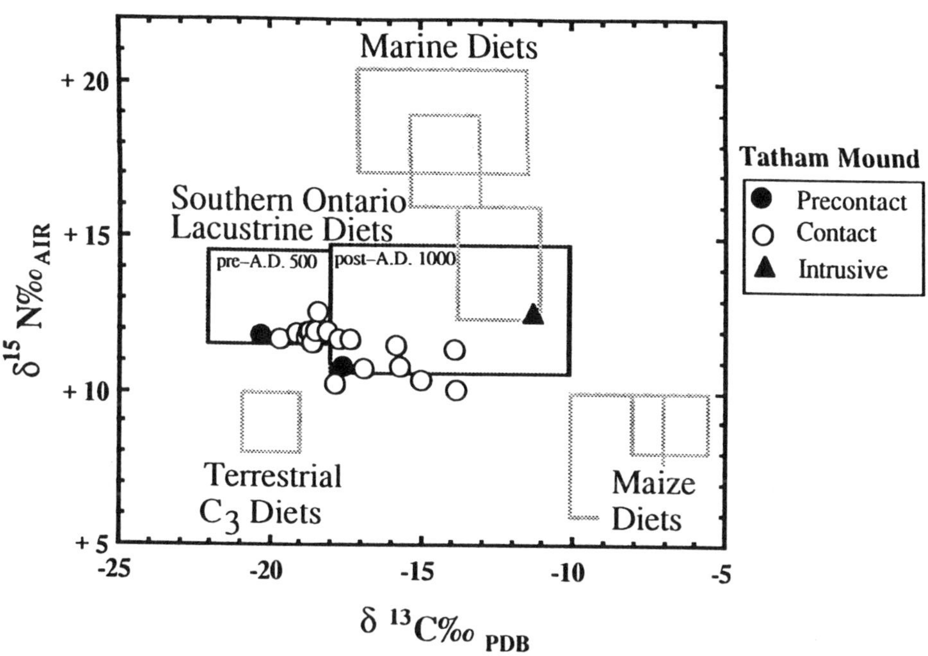

Fig. 7. Tatham Mound, Florida, compared to Ontario populations (boxes represent data from Schoeninger et al., 1983, and Schwarcz et al., 1985).

flora, the dietary interpretations of isotope results are preliminary. Additional modern and archaeological samples of food resources from Florida are currently being analyzed to refine dietary reconstruction. One other line of evidence, dental caries, provides limited evidence for maize as a dietary staple. Less than half of the individuals from Tatham Mound with teeth ($n = 11$) have carious lesions, and most of those cases consist of one affected tooth. We conclude that the stable carbon and nitrogen isotopic values from Tatham Mound are consistent with reports from the de Soto *entrada* indicating that maize was present only in limited amounts in the central Gulf coast region in the mid-1600s. The dietary stability at Tatham is in striking contrast to that in the Georgia Bight, where there was a reduction in dietary breadth and an increase in the consumption of maize after the establishment of mission centers on barrier islands (Larsen et al., 1992). These findings support ethnohistoric reports of diverse subsistence regimes within Florida during the early contact period.

ACKNOWLEDGMENTS

This research was funded by the Margaret Cullinan Wray Trust Fund. We thank Jerald T. Milanich of the Florida Museum of Natural History for providing archaeological samples. Bone samples were prepared for isotopic analysis at the University of Illinois Department of Anthropology Stable Isotope Lab, and mass spectrometry was performed at the Illinois State Geological Survey. We thank Stanley H. Ambrose for his comments on collagen extraction and Valerie Williams for laboratory assistance. For editing assistance and other comments we thank Lorraine V. Aragon, Laura Kozuch, William H. Marquardt, Lee Newsom, and Karen Walker.

REFERENCES CITED

Ambrose SH (1990) Preparation and characterization of bone and tooth collagen for isotopic analysis. Journal of Archaeological Science *17*: 431-451.

Ambrose SH (1993) Isotopic analysis of paleodiets: Methodological and interpretive considerations. In MK Sandford (ed.): Investigations of Ancient Human Tissue: Chemical Analyses in Anthropology. Philadelphia: Gordon and Breach, pp. 59–130.

Ambrose SH, and DeNiro MJ (1986) The isotopic ecology of east African mammals. Oecologia *69*:395–406.

Ambrose SH, and Norr L (n.d.) Experimental evidence for the relationship of the carbon isotope ratios of whole diet and dietary protein to those of bone collagen and carbonate. In J Lambert and G Grupe (eds.): Molecular Archaeology of Prehistoric Human Bone. Berlin: Springer-Verlag, in press.

Benedict CR, Wong WWL, and Wong JHH (1980) Fractionation of the stable isotopes of inorganic carbon by seagrasses. Plant Physiology *65*:512–517.

Bennett CE (trans.) (1975) Three Voyages. Gainesville: University of Florida Press.

Bourne EG (ed.) (1922) Narratives in the Career of Hernando de Soto. Two volumes. Translated by EG Bourne. New York: Allerton Book Company.

Chisholm BS, Nelson DE, and Schwarcz HP (1982) Stable-carbon isotope ratios as a measure of marine versus terrestrial protein in ancient diets. Science *216*:1131–1132.

Deines P (1980) The isotopic composition of reduced organic carbon. In P Fritz and JCh Fontes (eds.): Handbook of Environmental Isotope Geochemistry. Amsterdam: Elsevier, pp. 329–406.

DeNiro MJ (1985) Postmortem preservation and alteration of in vivo bone collagen isotope ratios in relation to paleodietary reconstruction. Nature *317*:806–809.

DeNiro MJ, and Epstein S (1978) Influence of diet on the distribution of carbon isotopes in animals. Geochimica et Cosmochimica Acta *42*: 495–506.

DeNiro MJ, and Epstein S (1981) Influence of diet on the distribution of nitrogen isotopes in animals. Geochimica et Cosmochimica Acta *45*: 341–351.

Fitzgerald C (1987) Analysis of the faunal remains at the Bayonet Field Site (8Ci197), Citrus County,

Florida. Manuscript on file, Florida Museum of Natural History, Gainesville.

Fontaneda E (1868) Memoria de las cosas y costa y indios de la Florida, que ninguno de cuantos la han costeado, no lo han sabido declarar. In: Colección de Documentos Inéditos, Relativos al Descubrimiento, Conquista y Organización de las Antiguas Posesiones Españolas de América y Oceania, vol. 5, pp. 532–548. 42 Volumes. Madrid: J. M. Perez.

Goggin JM, and Sturtevant WT (1964) The Calusa: A stratified nonagricultural society (with notes on sibling marriage). In W Goodenough (ed.): Explorations in Cultural Anthropology: Essays in Honor of George Peter Murdock. New York: McGraw-Hill, pp. 179–219.

Griffin JW (1988) The Archeology of the Everglades National Park: A Synthesis. Tallahassee: National Park Service Southeast Archeological Center.

Hann JH (1988) Apalachee: The Land Between the Rivers. Gainesville: University Presses of Florida.

Hann JH (1991) Missions to the Calusa. Gainesville: University Presses of Florida.

Hutchinson DL (1991) Postcontact Native American Health and Adaptation: Assessing the Impact of Introduced Diseases in Sixteenth-Century Gulf Coast Florida. Ph.D. dissertation, Department of Anthropology, University of Illinois, Urbana-Champaign.

Johnson WG (1990) The role of maize in south Florida aboriginal societies: An overview. Florida Anthropologist 43:209–214.

Keegan WF (1987) Diffusion of maize from South America: The Antillean connection reconstructed. In WF Keegan (ed.): Emergent Horticultural Economies of the Eastern Woodlands. Carbondale: Center for Archaeological Investigations, Occasional Paper 7, pp. 329–344.

Keegan WF, and DeNiro MJ (1988) Stable carbon- and nitrogen-isotope ratios of bone collagen used to study coral-reef and terrestrial components of prehistoric Bahamian diet. American Antiquity 53:320–336.

Kozuch L (1986) An overview of faunal lists from selected Safety Harbor sites. Manuscript on file, Department of Anthropology, Florida Museum of Natural History, Gainesville.

Larsen CS, Schoeninger MJ, van der Merwe NJ, Moore KM, and Lee-Thorp JA (1992) Carbon and nitrogen stable isotopic signatures of human dietary change in the Georgia Bight. American Journal of Physical Anthropology 89:197–214.

Lorant S (ed.) (1965) The New World: The First Pictures of America, rev. ed. New York: Duell, Sloan and Pearce.

Marquardt WH (1986) The development of cultural complexity in southwest Florida: Elements of a critique. Southeastern Archaeology 5:63–70.

Marquardt WH (1992) Culture and Environment in the Domain of the Calusa. Gainesville: University of Florida, Institute of Archaeology and Paleoenvironmental Studies.

Milanich JT (1987) Corn and Calusa: De Soto and Demography. In SW Gaines (ed.), Coasts, Plains, and Deserts: Essays in Honor of Reynold J. Ruppe. Arizona State University Anthropological Papers 38, pp 173–184.

Milanich JT, Chapman J, Cordell AS, Hale S, and Marrinan RA (1984) Prehistoric development of Calusa society in southwest Florida: Excavations on Useppa Island. In DD Davis (ed.): Perspectives on Gulf Coast Prehistory. Gainesville: University of Florida Press, pp. 258–314.

Milanich JT, and Fairbanks CH (1980) Florida Archaeology. New York: Academic Press.

Mitchem JM (1989a) Ethnohistoric and archaeological evidence for a protohistoric provincial boundary in west peninsular Florida. Paper presented, Southeastern Archaeological Conference, Tampa.

Mitchem JM (1989b) Redefining Safety Harbor: Late Prehistoric/Protohistoric Archaeology in West Peninsular Florida. Ph.D. dissertation, Department of Anthropology, University of Florida, Gainesville.

Newsom LA (1987) Analysis of botanical remains from Hontoon Island (8Vo202), Florida: 1980–1985 excavations. Florida Anthropologist 40: 47–84.

Newsom LA (1991a) Horr's island archaeobotanical research. In WH Marquardt (ed.): Final Report of Archaic and Glades Settlement and Subsistence Patterns. Gainesville: Florida Museum of Natural History.

Newsom LA (1991b) Archaeobotanical analysis of four features from the 1990 South End project at Fig Springs (8Co1), Columbia County, Florida. Report on file, Florida Museum of Natural History, Gainesville.

Newsom LA, and Quitmyer IR (1992) Archaeobotanical and Faunal Remains. In BR Weisman, Excavations on the Franciscan Frontier. Gainesville: University Presses of Florida, pp. 206–233.

Norr L (1991) Nutritional Consequences of Prehistoric Subsistence Strategies in Lower Central America. Ph.D. dissertation, Department of

Anthropology, University of Illinois, Urbana-Champaign.

Norr L (n.d.) Prehistoric coastal subsistence in northwestern Costa Rica: Geographical diversity and chronological trends. Revised version of a paper presented at the International Congress of Americanists, New Orleans, July 1991. In FW Lange (ed.): Weit ist der Weig? Papers in Honor of Wolfgang Haberland. Submitted.

Oviedo GF (1944) Historia General y Natural de las Idias: Islas y Tierra-Firme del Mar Oceano. 16 volumes. Asuncion del Paraguay: Editorial Guarania.

Scarry CM (1982) Paleoethnobotany of the Granada Site. In: Excavations at the Granada Site. Tallahassee: Florida Division of Archives, History and Records Management, pp. 181–248.

Scarry CM, and Newsom LA (1992) Archaeobotanical research in the Calusa heartland. In WH Marquardt (ed.): Culture and Environment in the Domain of the Calusa. Gainesville: Institute of Archaeology and Paleoenvironmental Studies, pp. 375–401.

Scarry JF (1990) Mississippian emergence in the Fort Walton area. In BD Smith (ed.): The Mississippian Emergence. Washington, DC: Smithsonian Institution Press, pp. 227–250.

Schoeninger MJ, and DeNiro MJ (1984) Nitrogen and carbon isotopic composition of bone collagen from marine and terrestrial animals. Geochimica et Cosmochimica Acta *48*:625–639.

Schoeninger MJ, DeNiro MJ, and Tauber H (1983) Stable nitrogen isotope ratios of bone collagen reflect marine and terrestrial components of prehistoric human diet. Science *220*:1381–1383.

Schwarcz HP, Melbye J, Katzenberg MA, and Knyf M (1985) Stable isotopes in human skeletons of southern Ontario: Reconstructing paleodiet. Journal of Archaeological Science *12*:187–206.

Sears E (1982) Pollen analysis. In WH Sears: Fort Center: An Archaeological Site in the Lake Okeechobee Basin. Gainesville: University Presses of Florida, pp. 118–129.

Sears EO, and Sears WH (1976) Preliminary report on prehistoric corn pollen from Fort Center, Florida. Southeastern Archaeological Conference Bulletin *19*:53–56.

Smith B (1966) Relation of Alvar Nuñez Cabeca de Vaca. Ann Arbor: March of America Facsimile Series 9.

Smith B (1968) Narratives in the Career of Hernando de Soto in the Conquest of Florida. Translated by Buckingham Smith. Reprinted. Gainesville: Kallman Publishing Company. Originally published 1866, Bradford Club, New York.

Smith BD (ed.) (1978) Mississippian Settlement Patterns. New York: Academic Press.

Smith BD (ed.) (1990) The Mississippian Emergence. Washington, DC: Smithsonian Institution Press.

Smith BN (1972) Natural abundance of the stable isotopes of carbon in biological systems. BioScience *22*:226–231.

Smith BN, and Epstein S (1971) Two categories of $^{13}C/^{12}C$ ratios for higher plants. Plant Physiology *47*:380–384.

Stark BL, and Voorhies B (eds.) (1978) Prehistoric Coastal Adaptations: The Economy and Ecology of Maritime Middle America. New York: Academic Press.

Stirling MW (1930) Prehistoric Mounds in the Vicinity of Tampa Bay, Florida. In: Explorations and Fieldwork of the Smithsonian Institution in 1929. Washington, DC: Government Printing Office, pp. 183–186.

Sturtevant WC (1962) Spanish-Indian relations in southeastern North America. Ethnohistory *9*: 41–94.

van der Merwe NJ (1982) Carbon isotopes, photosynthesis, and archaeology. American Scientist *70*:596–606.

van der Merwe NJ, and Vogel J (1978) ^{13}C content of human collagen as a measure of prehistoric diet in woodland North America. Nature *276*: 815–816.

Vogel J (1978) Isotopic assessment of the dietary habits of ungulates. South African Journal of Science *74*:298–301.

Walker KJ (1992) The zooarchaeology of Charlotte Harbor's prehistoric maritime adaptation: Spatial and temporal perspectives. In WH Marquardt (ed.): Culture and Environment in the Domain of the Calusa. Gainesville: Institute of Archaeology and Paleoenvionmental Studies, pp. 265–366.

Walker PL, and DeNiro MJ (1986) Stable nitrogen and carbon isotope ratios in bone collagen as indices of prehistoric dietary dependence on marine and terrestrial resources in southern California. American Journal of Physical Anthropology *71*:51–61.

Widmer RJ (1988) The Evolution of the Calusa: A Nonagricultural Chiefdom on the Southwest Coast of Florida. Tuscaloosa: University of Alabama Press.

Willey GR (1949) Archaeology of the Florida Gulf Coast. Smithsonian Miscellaneous Collections 113.

Chapter 3

The Stresses of Conquest in Spanish Florida: Structural Adaptation and Change Before and After Contact

Clark Spencer Larsen and Christopher B. Ruff

Research Laboratories of Anthropology and Department of Anthropology, University of North Carolina, Chapel Hill, NC 27599 (C.S.L.); Department of Anthropology, The American Museum of Natural History, New York, NY 10024 (C.S.L.); and Department of Cell Biology and Anatomy, School of Medicine, The Johns Hopkins University, Baltimore, MD 21205 (C.B.R.)

Beginning with the arrival of Europeans to the New World, native populations provided an indispensable labor source for many purposes, including construction projects, mining, agriculture, and transportation. Indeed, the economic success of Spanish colonizers was dependent on the extraction of labor from indigenous peoples. In *La Florida*—the vast region of eastern North America named by Ponce de León in 1513—written documentation provided by priests, explorers, and others shows that the colonists' arrival, especially the establishment of Catholic missions in northern Florida and coastal Georgia beginning in the late sixteenth century, occasioned dramatic behavioral changes in native populations. Elsewhere we have documented a number of health and work-related changes in postcontact and precontact populations in Guale, the name associated with the tribe and region encompassing the Atlantic coast and its barrier islands in present-day Georgia and northern Florida (reviewed in Larsen, 1993).

Eventually, changes wrought by Spanish contact led to the extinction of native populations in this part of Spanish Florida (Larsen et al., 1992). Although present knowledge of this process is far from complete, morphological and structural changes in long bones are an important means of documenting and understanding the new functional adaptations to physical behavior, particularly increased labor demands. Human remains from contact period sites have been found in all of the major provinces of Spanish Florida—Timucua, Apalachee, and Guale. Guale is unique, however, in that large collections of human remains from precontact sites provide an opportunity to contrast ways of life both before and after the arrival of Europeans in the region with respect to physical activity.

BIOCULTURAL SETTING

Prior to initial early sixteenth-century contact of Guale by Europeans, native populations exploited a wide range of plants and animals. Terrestrial resources contrib-

In the Wake of Contact: Biological Responses to Conquest, pages 21–34

uted significantly to the diet, but marine resources (e.g., fishes and shellfish) provided the majority of protein for much of prehistory (Reitz, 1988). During the last several centuries before contact—by at least the twelfth century A.D.—maize had become an important food source. Although archaeobotanical evidence for this dietary shift is scanty, stable nitrogen and carbon isotope ratios from human bone collagen clearly indicate a dietary reorientation toward terrestrial resources (Larsen et al., 1992). This shift in diet is consistent with the appearance of large and more complex societies, increased sedentism, and greater population sizes in many parts of the eastern United States (Smith, 1989; Anderson, 1990).

When Spaniards first arrived on the Georgia coast, they undoubtedly recognized the enormous potential of native crops and other dietary resources for the long-term support of a European presence. In addition, it was soon realized that Indian labor could be used for planting, maintaining, and harvesting economically viable crops. Written documentation strongly suggests that the Spanish were dependent on this cheap source of labor. For instance, Governor Canzo reported to the Spanish crown that

> but with all this and the grain from maize, the labor that they endure in the many cultivations that are given is great, and, if it were not for the help of the Indians that I make them give . . . it would not be possible to be able to sow any grain. . . . (translation by John H. Hann; cited in Larsen, 1990)

Other accounts also note the use of Indian laborers for bearing cargo, building projects, cutting wood, and other physically taxing activities, in addition to agricultural tasks (Hann, 1988).

Given Spanish needs, it seems quite plausible that work demands, particularly pertaining to agricultural production, increased after contact. Indians were not only responsible for feeding themselves, but also for providing food for Spanish mission priests and military personnel and for export to other parts of *La Florida*, including the capital at St. Augustine. In order to control this labor source, as well as to facilitate religious conversion and acculturation, the Spanish followed a policy of *reducción* whereby native peoples were packed into settlements around mission centers. These forced settlement changes contributed to increased sedentism, especially in comparison to prehistoric ways of life (Hann, 1988).

GUALE SKELETAL REMAINS

The concomitant effects of increased work demands and sedentism on native populations can be examined by dividing precontact and contact human skeletons into four groups representing successive time periods: *precontact preagricultural* (pre-A.D. 1150), *precontact agricultural* (A.D. 1150–1550), *early contact* (A.D. 1607–1680), and *late contact* (A.D. 1686–1702). Native peoples during the first period were exclusively hunter-gatherer-fishers who relied upon wild plants and animals from both marine and terrestrial settings. Human remains from this period come from various Georgia coastal localities (Larsen, 1982). During the precontact agricultural period, groups continued to use wild plants and animals, but maize also became an important component of the diet. Human remains for this period are drawn largely from the Irene Mound site, the dominant Mississippian period locality in the region (Caldwell and McCann, 1941; Larsen, 1982).

During the late sixteenth century, several attempts were made to establish permanent Catholic missions on the Georgia coast. Not until the early seventeenth century, however, were the missions to achieve any semblance of stability (Jones, 1978). Santa Catalina de Guale on St. Catherines Island, Georgia, served as the principal mission on the Georgia coast for most of the 1600s until the British and their Indian allies forced its abandonment in 1680. By 1686, the mission was reestablished south of the St. Marys River on Amelia Island, Florida. It was again destroyed in 1702 by the British along with all other missions north of St. Augustine. Historical documents, stable isotopic analyses of bones, and dental caries frequencies demonstrate that these populations depended on maize to a greater extent than their prehistoric forebears (Larsen et al., 1992). The early contact period is represented by human remains from the Santa Catalina mission on St. Catherines Island as well as several individuals from the nearby contemporary Pine Harbor site, and the late contact period is represented by human remains from the later Santa Catalina mission on Amelia Island (see descriptions in Larsen [1982, 1990, 1993]).

LONG BONE STRUCTURAL ANALYSIS

Anthropometric approaches to skeletal metrics have held a prominent place in osteological studies since the nineteenth century. Unfortunately, the imprecision of these measurements as they relate to behavioral patterns limits their value in learning about how people once lived (Larsen, 1987). In recent years, the application of beam theory developed by civil and mechanical engineers to the analysis of long bone diaphyses has added an important means of reconstructing the lifestyles of past populations (Ruff, 1992). Like materials used in man-made buildings, long bones of upper and lower limbs (e.g., humerus and femur) can be modeled as hollow beams and the "strength" (more precisely, rigidity) of these beams can be measured directly. The application of such a model facilitates the summary of complex long bone shapes into a series of readily interpretable properties—called cross-sectional geometric properties—allowing functional and behavioral inferences.

Femora and humeri from the four Guale groups were oriented in standard anteroposterior and mediolateral planes and cut transversely at the midshaft and lesser trochanteric regions of femoral diaphyses and in the mid-distal region of the humeral diaphyses (Ruff and Hayes, 1983). Exposed sections were photographed, and rear-projected images of the inner (endosteal) and outer (periosteal) surfaces were electronically traced on a digitizer screen. Calculations of section properties were made using a microcomputer interfaced with the digitizer apparatus.

Two types of properties representing bone strength—"areas" and "second moments of area"—were calculated by a modified version of the computer program SLICE used in structural analyses (Nagurka and Hayes, 1980; Ruff, 1992). Areas reflect the amount of bone in a cross section, and include cortical area (CA), medullary area (MA), and total subperiosteal area (TA). Second moments of area—maximum and minimum (I_{max}, I_{min}), mediolateral and anteroposterior (I_x, I_y), and polar (J) second moments of area—reflect the distribution of bone in relation to axes running through the section. I and J values represent bending and torsional (twisting) strength, respectively, the two primary loading modes affecting long bones. Simply stated, a decrease in properties represents a decrease in bone strength; that is, the ability to resist mechanical loading. Conversely, an increase represents a greater ability to resist mechanical loading.

TEMPORAL TRENDS IN LONG BONE SHAPE

Trends in bone lengths and cross-sectional geometric properties for male and female femora and humeri point to a number of temporal changes in Guale (Table 1). Overall, lengths of male long bones show relatively less change than lengths of their female counterparts. For example, females show a marked decrease in bone length followed by less dramatic changes in early con-

TABLE 1. Bone Size and Cross-sectional Geometric Property Summary Statistics: Precontact Preagricultural (PP), Precontact Agricultural (PA), Early Contact (EC), and Late Contact (LC)[a]

Property[b]	PP		PA		EC		LC		Significant differences[c]
	Mean	SE	Mean	SE	Mean	SE	Mean	SE	
Male									
Femur	($n=8$)		($n=11$)		($n=9$)		($n=24$)		
Length′	424.9	7.9	411.2	3.7	420.3	7.9	420.0	3.9	PP/PA
Midshaft									
CA/$\ln^2$	233.6	14.0	234.5	7.4	237.8	11.6	224.5	4.5	
MA/$\ln^2$	82.2	8.1	61.8	4.0	89.6	5.9	97.8	4.2	PP/PA,PA/EC,PA/LC
TA/$\ln^2$	315.8	12.5	296.3	7.7	307.9	15.6	322.3	5.7	PA/LC
$I_{max}/\ln^4$	88.6	8.3	77.9	5.0	89.1	5.7	92.4	3.9	PA/LC
$I_{min}/\ln^4$	65.5	4.8	61.0	2.9	63.8	4.5	65.7	2.3	
$I_x/\ln^4$	86.8	8.1	71.7	4.0	81.6	5.5	81.1	2.7	PP/PA,PA/LC
$I_y/\ln^4$	67.3	4.8	67.2	4.4	71.3	5.8	73.4	3.4	
$J/\ln^4$	154.1	12.6	138.9	7.5	152.9	8.9	158.1	5.4	PA/LC
I_x/I_y	1.28	—	1.08	—	1.15	—	1.11	—	PP/PA,PP/LC
Subtrochanteric									
CA/$\ln^2$	225.0	10.4	220.8	7.1	236.2	8.5	225.1	5.1	
MA/$\ln^2$	123.6	11.0	83.6	6.1	113.1	6.3	123.0	5.2	PP/PA,PA/EC,PA/LC
TA/$\ln^2$	348.6	15.7	304.3	9.0	349.3	11.2	348.1	5.4	PP/PA,PA/EC
$I_{max}/\ln^4$	124.6	12.2	92.6	5.2	110.8	7.0	107.6	4.0	PP/PA,PA/EC
$I_min/\ln^4$	59.1	4.4	51.5	3.4	69.0	4.9	66.6	2.0	PA/EC,PA/LC
$J/\ln^4$	183.7	16.2	144.1	8.3	179.9	11.3	174.3	5.5	PP/PA,PA/EC,PA/LC
I_{max}/I_{min}	2.09	—	1.82	—	1.61	—	1.62	—	PA/EC,PA/LC
Humerus	($n=15$)		($n=15$)		($n=11$)		($n=24$)		
Length′	318.3	5.1	312.5	2.6	313.6	13.4	313.5	2.7	
Mid-distal									
CA/$\ln^2$	215.0	11.2	201.2	6.9	204.8	10.2	215.7	5.0	
MA/$\ln^2$	83.2	9.5	70.9	7.8	81.8	7.4	100.0	6.8	PA/LC
TA/$\ln^2$	298.2	11.8	272.1	6.7	286.6	10.7	315.7	6.1	PP/PA,PA/LC
$I_{max}/\ln^4$	73.8	5.3	62.8	2.9	69.5	5.3	77.7	2.8	PP/PA,PA/LC
$I_{min}/\ln^4$	61.1	4.8	48.8	2.7	55.5	5.7	64.9	2.5	PP/PA,PA/LC
$I_x/\ln^4$	66.4	4.8	56.0	2.9	64.3	4.7	73.5	2.7	PP/PA,PA/LC
$I_y/\ln^4$	68.6	5.5	55.5	2.4	60.7	6.4	69.1	2.9	PP/PA,PA/LC
$J/\ln^4$	135.0	10.1	111.6	5.1	125.0	10.9	142.6	5.2	PP/PA,PA/LC
I_x/I_y	.98	—	1.01	—	1.06	—	1.06	—	PP/EC,PP/LC,PA/EC

Female									
Femur	(*n* = 12)		(*n* = 9)		(*n* = 9)		(*n* = 23)		
Length′	415.2	6.0	376.4	5.0	404.1	6.5	392.1	3.2	PP/PA,PP/LC,PA/EC,PA/LC
Midshaft									
CA/ln^2	190.2	10.8	186.4	7.8	210.0	6.6	204.8	7.7	PA/EC
MA/ln^2	93.2	9.1	70.5	5.0	74.1	7.0	100.4	6.7	PP/PA,PA/LC,EC/LC
TA/ln^2	283.5	5.9	256.9	10.1	284.0	7.1	305.2	7.4	PP/PA,PA/EC,PA/LC
I_{max}/ln^4	63.4	3.7	54.8	4.2	67.5	2.9	75.3	4.0	PA/EC,PA/LC
I_{min}/ln^4	51.9	2.5	44.9	3.6	54.0	2.9	65.7	7.6	PA/EC
I_x/ln^4	62.0	3.9	50.6	5.0	60.0	3.0	65.7	3.9	PP/PA,PA/EC,PA/LC
I_y/ln^4	53.4	4.6	49.2	4.2	61.6	2.8	69.9	3.7	PP/PA,PP/LC,PA/EC,PA/LC
J/ln^4	115.4	5.9	99.8	7.8	121.6	5.3	141.0	11.1	PA/EC,PA/LC
I_x/I_y	1.16	—	1.03	—	.97	—	.94	—	PP/LC
Subtrochanteric									
CA/ln^2	191.5	10.8	190.4	5.7	215.8	7.6	206.6	6.9	PA/EC
MA/ln^2	120.2	12.7	88.6	6.0	95.9	9.0	129.2	7.7	PP/PA,PA/LC,EC/LC
TA/ln^2	311.7	6.9	279.0	9.2	311.7	13.1	335.8	8.1	PP/PA,PA/EC,PA/LC
I_{max}/ln^4	94.7	4.7	81.8	5.9	94.7	7.8	100.7	4.9	PA/LC
I_{min}/ln^4	44.9	2.5	39.2	2.3	52.9	4.3	58.6	3.1	PP/EC,PP/LC,PA/EC,PA/LC
J/ln^4	139.6	6.6	121.0	7.8	147.6	11.5	159.3	7.5	PP/PA,PA/EC,PA/LC
I_{max}/I_{min}	2.14	—	2.10	—	1.79	—	1.72	—	PP/EC,PP/LC,PA/EC,PA/LC
Humerus	(*n* = 15)		(*n* = 15)		(*n* = 11)		(*n* = 24)		
Length′	305.8	7.8	290.2	2.8	301.1	5.6	298.6	2.6	PP/PA,PA/EC,PA/LC
Mid-distal									
CA/ln^2	179.4	10.8	173.2	7.7	169.7	6.6	161.3	6.2	
MA/ln^2	72.5	8.9	71.7	7.4	58.2	4.5	94.9	6.7	PA/LC,EC/LC
TA/ln^2	251.9	10.7	244.9	7.0	227.9	7.4	256.1	5.9	EC/LC
I_{max}/ln^4	53.9	4.3	50.8	2.7	45.9	3.1	52.2	2.4	
I_{min}/ln^4	40.7	3.6	38.4	2.6	34.3	2.6	38.8	2.1	
I_x/ln^4	50.4	4.5	45.8	2.7	42.3	3.3	47.4	2.4	
I_y/ln^4	44.3	3.4	43.4	2.6	37.5	2.4	43.6	2.3	
J/ln^4	94.6	7.8	89.2	5.1	80.1	5.6	91.0	4.5	
I_x/I_y	1.13	—	1.06	—	1.13	—	1.09	—	

(continued)

TABLE 1. *(Continued)*

[a]From Ruff et al., 1984; Ruff and Larsen, 1990; present study.

[b]Explanation of properties: femur length′ (in mm) is the distance measured from the condyles to the superior margin of the neck, and humerus length′ is the distance from the head to the lip of the trochlea; midshaft and subtrochanteric femur refer to 50% and 80%, respectively, of length′ measured from the distal end; mid-distal humerus refers to 35% of bone length′ measured from distal end; area indices over bone length (CA, MA, TA) are multiplied by 10^5, and second moment of area indices over bone length (I_{max}, I_{min}, I_x, I_y, J) are multiplied by 10^8; CA, cortical area; MA, medullary area; TA, total subperiosteal area; I_{max}, maximum second moment of area (bending strength); I_{min}, minimum second moment of area (bending strength); I_x, second moment of area about mediolateral axis (anteroposterior bending strength); I_y, second moment of area about anteroposterior axis (mediolateral bending strength); J, polar second moment of area. Humeral properties are averaged for right and left sides, when available, or values adjusted for average bilateral differences (see Ruff and Larsen, 1990).

Area indices (CA, MA, TA) provide relative measures of the *amount* of cortical bone in a cross section and the relative size of the section and medullary cavity; I (bending strength) and J (torsional strength) reflect the *distribution* of bone about a neutral axis or point; ratios (I_x/I_y, I_{max}/I_{min}) reflect relative distribution of bone about perpendicular axes, and thus cross-sectional "shape" (see Ruff and Hayes, 1983; Ruff et al., 1984; Ruff and Larsen, 1990).

All properties are standardized for body size differences between both periods and sexes by dividing areas by bone length^2 and second moments of area by bone length^4 (see Ruff, 1984).

[c]$p \leq 0.05$, *t* test (two-tailed).

tact and late contact samples. These data show that females experienced a considerable body size reduction in the precontact agricultural period, which rebounded in the early contact period and was followed by a slight decline later in time. Males exhibit virtually no change in body size between the early and late contact periods.

Cross-sectional geometric properties also exhibit a number of trends over time. Precontact preagricultural and agricultural femora show no appreciable change in CA. However, both MA and TA for both sexes decline markedly, thus reflecting a contraction of endosteal and periosteal surfaces. In sharp contrast to the precontact trend, the early and late contact periods show noticeable increases in MA and TA. With regard to the second moments of area, the general trend is a decrease in the precontact agricultural period followed by successive increases in the two contact periods. This trend occurs without exception for the female properties. In males, the overall direction is less clear. Although all properties increase in the early contact period, several properties decline slightly (e.g., subtrochanteric I_{max}) or level off (midshaft I_x) in the late contact period.

The humeri for both sexes show the same basic temporal pattern of structural change. In males, all properties—areas and second moments of area—first decline prior to contact, then increase in both contact periods. In contrast to males, female properties continue to decline in the early contact period, but this trend is reversed later in time, except for CA. Thus, both males and females show general increases in areas and second moments of area in the last period.

A representation of cross-sectional "shape" of long bones can be obtained from the ratio of I_x to I_y in the femur midshaft and the mid-distal humerus (Ruff, 1987). In the male femur midshaft, the I_x/I_y ratio measuring A-P/M-L bending strength decreases prior to contact, increases during early contact, and finally decreases once again. In the mid-distal humerus, there is a decrease before contact followed by an increase early in the contact period. No change is discernible later in the contact period. In contrast to males, in females there is a continuous decline in the femoral midshaft I_x/I_y index from precontact preagricultural to late contact samples. In the humerus, a decline occurs prior to contact, and is followed by an increase and then decrease in the early and late contact periods, respectively.

Lastly, ratios of I_{max} to I_{min} in the subtrochanteric region of the femur—representing a ratio of maximum to minimum bending strength—show declines in both sexes. Such ratios, however, are quite similar for the early and late contact males. These findings indicate that the femoral subtrochanteric region becomes more circular from the earliest to latest periods.

IMPLICATIONS OF LONG BONE STRUCTURAL CHANGES IN GUALE

These findings can be summarized into three general trends for the precontact and contact Guale. First, after an initial decline in body size with the adoption of maize agriculture, body size increased in the early contact period. After the move to Amelia Island, females became smaller, while males remained essentially the same. Second, there was a reduction in amount of bone cortex and relative bone strength before contact. Following the establishment of missions in the region, this trend reverses itself with increases in both the amount of bone and bone strength. Third, the femur shows an increased circularity and the humerus shows a decreased circularity in cross-sectional shape.

Bone morphology is influenced by various factors, both intrinsic (genetic) and extrinsic (environmental). With regard to the temporally proximate native populations of this region, it is unlikely that these skeletal changes are due to influences of genetics. Both cultural (Saunders, 1992) and biological (Griffin, 1993) data argue for population continuity across the prehistoric to historic interface. Thus, it is unlikely that structural changes in Guale people were a function of the in-migration of a more robust population in the contact period. It therefore seems doubtful that gene flow could explain the temporal trends in cross-sectional geometric variation.

We suggest that these changes in bone dimensions likely represent a complex interplay of nutritional and mechanical factors affecting these populations. We have discussed elsewhere the probability that the reduction in adult body size prior to contact was due to a general decline in nutritional quality in part having to do with a reliance on maize, which is lacking in some essential nutrients (Ruff et al., 1984). Therefore, increases in body size after initial contact could represent an improvement in nutrition, perhaps in terms of the quantity of food consumed, especially for females. Studies of stable isotope ratios and dental caries frequencies, however, argue against an improvement in dietary *quality* (Larsen et al., 1992). Moreover, isotopic evidence shows a decrease in marine resource consumption, which when coupled with an increase in carbohydrate consumption indicates a decline in resource diversity in the contact period (see Larsen, 1990). Finally, written sources from the contact period reveal that food shortages followed epidemics or Spanish retaliation for native revolts, which indicate events disrupting food supply (Jones, 1978).

The study of mobility in these populations might provide some insight into mechanical adaptation. Ruff (1987) has argued that measurement of A-P relative to M-L bending strength in the femur midshaft (I_x/I_y) is useful for estimating the relative degree of mobility, such as that related to long-distance travel or running. In general, higher values of I_x/I_y reflecting greater A-P bending strength represent greater mobility. Ruff (1987) found that in highly mobile hunter-gatherers, sedentary prehistoric agriculturalists, and highly sedentary industrial Western populations, relative A-P bending strength was highest in hunter-gatherers and lowest in industrial groups. Interestingly, this finding corresponds well with the increase in circularity of long bone external diaphyses from prehistoric hunter-gatherers to agriculturalists (e.g., Larsen, 1982).

The analysis of cross-sectional shape in the early contact sample reveals that although on average males were less mobile, at least some males became more mobile relative to precontact agriculturalists. It is known that Indian males were conscripted frequently into labor projects involving long-distance travel as part of the *Repartimiento* labor system (Ruff and Larsen, 1990). Therefore, we surmise that males showing skeletal changes indicative of higher mobility may have been engaged in travel to distant places. In contrast, all females show skeletal evidence for a decrease in mobility. On the other hand, both sexes experienced a decline in subtrochanteric femoral circularity that has been interpreted in other populations as reflecting a decline in *general* activity levels (see Ruff and Larsen, 1990).

Therefore, these findings indicate that the contact period populations were engaged in a combination of behaviors involving less mobility and reduced activity overall. Presumably, these behavioral modifications were related to the increasing sedentism characteristic of mission life and the *reducción* strategy employed by the Spanish to control native populations.

Explanation of the contact period increases in the magnitude of cross-sectional geometric properties (versus ratios) are more problematical. Perhaps body weight relative to stature was altered in some systematic fashion in the early and late contact groups.

A change in mobility and activity levels likely affected body weight, especially in populations relying on carbohydrates as a dietary mainstay. Mechanical influences on legs are related to the combined effects of body weight and physical activity. Thus, mechanical loading on legs should be proportional to body weight multiplied by a factor related to activity level. Here we have assumed that the body weight to bone length proportion remained constant when "size" standardizing cross-sectional geometric properties by dividing by second and fourth powers of length (areas and second moments of area, respectively; see Table 1 and Ruff and Larsen [1990] for explanation). It is conceivable, however, that the proportion of body weight to stature increased, because of greater sedentism and the effects of dietary change. If there was a contact period weight gain due to increasing sedentism, confinement of body movement, and carbohydrate consumption, then the body weight/bone length ratio and the length standardized properties would be increased. Moreover, it has been shown that the morphology of the femoral diaphysis is sensitive to changes in body weight, particularly during adulthood (see Ruff et al., 1991).

Because of the nature of the data, it is difficult to test the increased body weight hypothesis. Nevertheless, we suggest that a weight-related interpretation is consistent with trends observed in modern Native American groups in North America experiencing transitions to more sedentary lifestyles during the nineteenth and twentieth centuries. Hrdlička (1908) and Johnston and Schell (1979) have observed that sedentary groups consuming high-carbohydrate diets show a strong tendency for weight gain and obesity. Although these observations were made with reference to groups living in reservation and urban contexts (versus early missions), they provide a perspective on weight gain that may explain some of the observed geometric variation. In fact, at least two seventeenth-century observers commented on body morphology in *La Florida*. In the 1630 Memorial of Fray Francisco Alonso de Jesus, Indians were described as "corpulent" (Hann, 1992). Bishop Calderón remarked in 1675 that "they [the Indians] are fleshy, and rarely is there a small one" (cited in Hann, 1988:158). We strongly suspect that these observations report on the effects on body composition of alterations in settlement and diet during the contact period.

In contrast to the femur, the humerus shows a different pattern of structural change. In the female humerus, in particular, there is a continued reduction in bone strength through the early contact period. Because the humerus is subject only to forces derived from activities related to use of the upper limb—not body weight or ambulatory functions—it seems that the arm was exposed to a reduction in mechanical demand. There is no clear explanation for this reduction in mission females, except to speculate that women used their arms in activities requiring less physical exertion at the St. Catherines Island mission when compared to their precontact counterparts.

The late contact period females and early and late contact period males experienced an increase in humeral bone strength. These developments indicate that late contact females and contact period males increased the use of their upper limbs. The Spaniards used Indians for various forms of labor requiring demanding physical activity, including heavy lifting and carrying. Owing to the absence of good roads and the limited number of draft animals, Indians were frequently used to transport heavy materials, sometimes over great distances (Hann, 1988). Indians were also regularly employed for construction projects, such as building churches and secular structures. For example, the construction of the primary fortification at St. Augustine likely involved the participation of Guale Indians under strenuous circumstances over a period of years (Arana and Manucy, 1977). Historical sources do not provide a basis upon which to explain entirely changes in upper limb bone strength. Nevertheless, changes in bone structure indicate that physical demands on the upper limb increased for both females and males. Thus, both sexes may have been engaged in physical activities involving comparable loading modes.

Age composition of the skeletal samples used in the present analysis may be another important factor influencing the patterns of structural changes. Ruff and Hayes (1983) have observed increases in TA and second moments of area with an increase in age in the New Mexico Pecos Pueblo series, apparently as compensatory responses to endosteal bone loss during later adulthood. In the Georgia and Florida series, mean age-at-death decreases in the precontact agricultural period (from 26.1 years to 19.7 years) and increases both in the early contact (to 21.2 years) and late contact periods (to 28.8 years) (see Larsen et al. [1990] for a fuller consideration of Guale demographic change). The differences in sample composition could contribute to the temporal trends in cross-sectional areas and second moments of area.

At least for the precontact comparisons (preagricultural versus agricultural), it is unlikely that age-at-death explains the changes in cross-sectional geometric properties because it was possible to select individuals of similar ages from the two series (Ruff et al., 1984). Sample stratification by age was not possible in the two contact groups because both series contain few young adults. Examination of means of properties in Table 1, however, suggests that although age may have contributed to the trends, behavioral (and perhaps nutritional) factors were more important. For example, despite an eight-year increase in mean age-at-death for females represented by humeri in the early contact sample relative to the precontact agricultural sample, all areas decrease (Table 2). This finding supports our contention that although age increases in the sequence, other factors—most likely behavioral—are of far more importance in interpreting temporal trends in cross-sectional geometric properties. In a similar vein, late contact male femora are from older adults than precontact preagricultural male femora; however, the geometric properties are similar. None of the areas or second moments of area are statistically distinguishable between early and late contact in males (t test, $p > 0.05$) (Table 1).

TABLE 2. Mean Age-at-Death of Individuals by Skeletal Element: Precontact Preagricultural (PP), Precontact Agricultural (PA), Early Contact (EC), and Late Contact (LC)

Skeletal element	Period	Sex	Mean age
Femur	PP	Female	30.3
Femur	PP	Male	25.5
Femur	PA	Female	26.1
Femur	PA	Male	23.8
Femur	EC	Female	35.9
Femur	EC	Male	32.5
Femur	LC	Female	42.1
Femur	LC	Male	38.1
Humerus	PP	Female	29.8
Humerus	PP	Male	29.6
Humerus	PA	Female	29.0
Humerus	PA	Male	26.8
Humerus	EC	Female	37.0
Humerus	EC	Male	31.9
Humerus	LC	Female	42.1
Humerus	LC	Male	38.1

We do not argue that mechanical factors are solely responsible for the structural modifications observed in long bone cross-sections reported here. Yet, it is true that standardized values of bone mass or volume (CA) remain essentially unchanged regardless of temporal period or sex (Table 1). In contrast to this finding, properties used to measure the distribution of bone in cross section show marked changes. The overall pattern is consistent with adaptations to *localized*—that is, mechanical—factors rather than *systemic* (nutritional) factors, although the two certainly appear to be interrelated (see Ruff et al., 1984).

THE MECHANICAL ENVIRONMENT IN GUALE: PATHOLOGY

Other data support our model of dramatic alterations in behavior over time. Analysis of prevalence of osteoarthritis, a disorder arising from the "wear and tear" on articular joints due to activity, reveals that skeletal changes (mostly proliferative lipping) in the major weight-bearing and non-weight-bearing joints declined in the precontact agricultural period but increased dramatically in the late contact period (early contact period data are unavailable) (Fig. 1). Mirroring the changes observed in cross-sectional geometric properties, temporal trends in osteoarthritis suggest that articular joints were subject to a reduction in mechanical loading prior to contact. After contact, however, there was an increase in physical demands. Moreover, osteoarthritis differences between the sexes decreased in the late contact period, a finding suggested by mid-distal humerus cross-sectional properties.

Another feature observed in the late contact skeletal remains frequently associated with high levels of activity in human populations is spondylolysis, which is the separation of the neural arch of lumbar vertebrae—especially involving the fifth lumbar vertebra—owing to gradual fatigue or stress fracture. Five of 56 (8.9%) of individuals from the Santa Catalina mission on Amelia Island have the lumbar vertebral defect. Although this prevalence is not unusually high when compared with other archaeological series (see Gunness-Hay, 1980), it is greater than frequencies observed in all other Georgia coastal skeletal samples observed (< 5%; Larsen, unpublished data).

Various factors appear to be related to spondylolysis (e.g., hereditary disposition [Wiltse et al., 1975]). However, mechanical factors appear to figure most prominently in its explanation (Bridges, 1989; Merbs, 1989). Indeed, high prevalence of spondylolysis in human groups following physically demanding lifestyles argues for a mechanical cause (e.g., Eskimos [Stewart, 1931, 1979; Lester and Shapiro, 1968; Merbs, 1983]). Moreover, spondylolysis appears to be related to specific activity types. For example, athletes tend to show very high prevalence, particularly heavier individuals with relatively high muscular strength in the back (e.g., Hoshina, 1980). Based on specific patterns shown in archaeological skeletons and data presented in the sports medicine literature, bending or twisting of the lower spine or heavy lifting may be more important factors than high activity levels per se (Bridges, 1989; Merbs, 1989). Thus, an increase in prevalence in the Spanish Florida series is consistent with a population undergoing increased physical demands.

Like cross-sectional changes, spondylolysis may be an adaptive response to changing mechanical circumstances. Merbs (1989) noted that the defect and accompanying changes to the spine may add greater flexibility to the lower back, thus enabling the individual to experience a reduced degree of discomfort while undertaking physically demanding activities.

CONCLUSION

The investigation of mechanical adaptation offers important insights into behavioral modifications associated with major adaptive shifts and generalized societal disruption after the establishment of missions in Guale. Both prior to and during the period of missionization and sustained contact with Europeans, native populations experienced marked alterations in settlement, diet, and habitual activity affecting bone morphology. In particular, the founding of missions among the Guale resulted in a shift to a more sedentary way of life with accom-

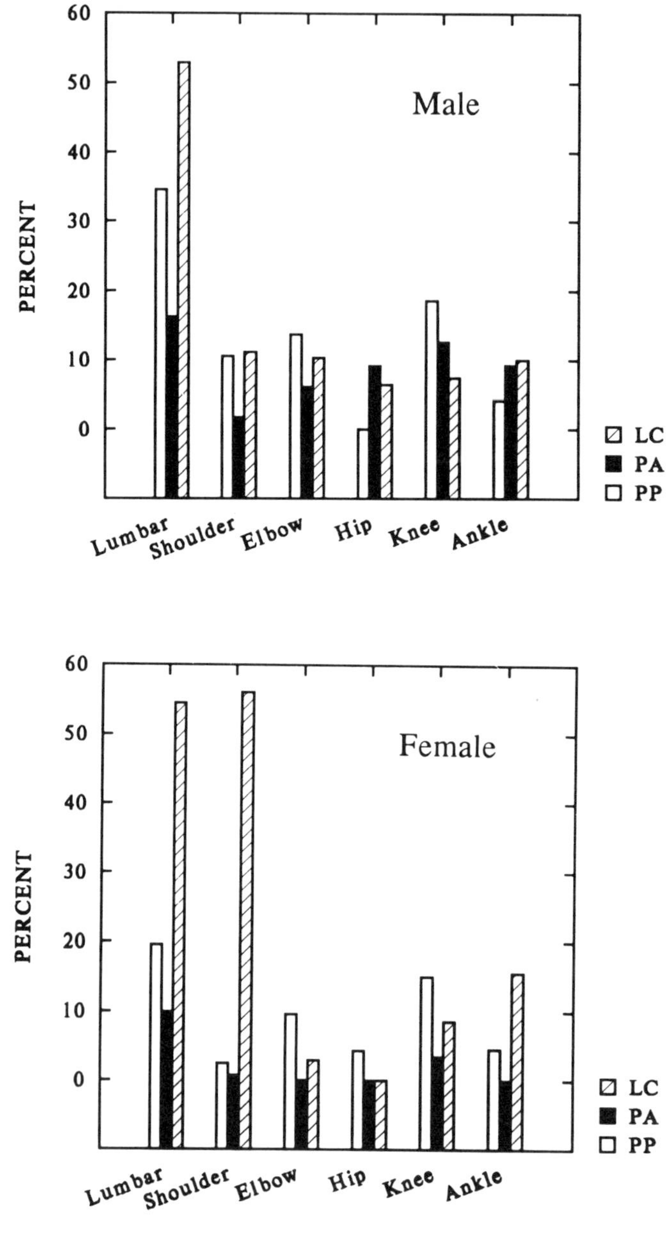

Fig. 1. Adult male and female osteoarthritis frequencies by joint: precontact preagricultural (PP), precontact agricultural (PA), and late contact (LC) (data from Griffin and Larsen, 1989; Larsen and Ruff, 1991).

panying decreases in general levels of activity and mobility during the seventeenth century. New circumstances and demands related to missionization were likely important contributing factors increasing body size and robusticity.

ACKNOWLEDGMENTS

We thank the Edward John Noble Foundation, the St. Catherines Island Foundation, and Dr. and Mrs. George H. Dorion for funding the fieldwork. The National Science Foundation (BNS-8406773, BNS-8703849, BNS-8747309) and Sigma Xi funded the analysis. Douglas Ubelaker and Donald Ortner gave permission for the study of human remains in the collections of the National Museum of Natural History, Smithsonian Institution.

REFERENCES CITED

Anderson DG (1990) Political Change in Chiefdom Societies: Cycling in the Late Prehistoric Southeastern United States. Ph.D. dissertation, Department of Anthropology, University of Michigan, Ann Arbor.

Arana LR, and Manucy A (1977) The Building of Castillo de San Marcos. Washington, DC: National Park Service.

Bridges PS (1989) Spondylolysis and its relationship to degenerative joint disease in the prehistoric southeastern United States. American Journal of Physical Anthropology 79:321–329.

Caldwell JR, and McCann C (1941) Irene Mound Site. Athens: University of Georgia Press.

Griffin MC (1993) Morphological Variation of the Late Precontact and Contact Period Guale. Ph.D. dissertation, Department of Sociology and Anthropology, Purdue University, West Lafayette.

Griffin MC, and Larsen CS (1989) Patterns in osteoarthritis: A case study from the prehistoric and historic southeastern U.S. Atlantic coast. American Journal of Physical Anthropology 78:232.

Gunness-Hey M (1980) The Koniag Eskimo presacral vertebral column: Variations, anomalies and pathologies. Ossa 7:99–118.

Hann JH (1988) Apalachee: The Land Between the Rivers. Gainesville: University Presses of Florida.

Hann JH (1992) Translation of 1630 memorial of Fray Francisco Alonso de Jesus. Manuscript on file, San Luis Archaeological and Historic Site, Florida Bureau of Archaeological Research, Tallahassee.

Hoshina H (1980) Spondylolysis in athletes. Physician and Sportsmedicine 8(9):75–79.

Hrdlička A (1908) Physiological and Medical Observations Among the Indians of the Southwestern United States and Northern Mexico. Bureau of American Ethnology Bulletin 34.

Johnston FE, and Schell LM (1979) Anthropometric variation of Native American children and adults. In WS Laughlin and AB Harper (eds.): The First Americans: Origins, Affinities, and Adaptations. New York: Gustav Fischer, pp. 275–291.

Jones GD (1978) The ethnohistory of the Guale coast through 1684. In DH Thomas, GD Jones, RS Durham, and CS Larsen: The Anthropology of St. Catherines Island: 1. Natural and Cultural History. Anthropological Papers of the American Museum of Natural History 55:178–210.

Larsen CS (1982) The Anthropology of St. Catherines Island: 3. Prehistoric Human Biological Adaptation. Anthropological Papers of the American Museum of Natural History 57:155–270.

Larsen CS (1987) Bioarchaeological interpretations of subsistence economy and behavior from human skeletal remains. In MB Schiffer (ed.): Advances in Archaeological Method and Theory, vol. 10. San Diego: Academic Press, pp. 339–445.

Larsen CS (1990) Biocultural interpretation and the context for contact. In CS Larsen (ed.): The Archaeology of Mission Santa Catalina de Guale: 2. Biocultural Interpretations of a Population in Transition. Anthropological Papers of the American Museum of Natural History 68:11–25.

Larsen CS (1993) On the frontier of contact: Mission bioarchaeology in *La Florida*. In BG McEwan (ed.): The Missions of Spanish Florida. Gainesville: University Presses of Florida, pp. 322–356.

Larsen CS, and Ruff CB (1991) Behavioral adaptations at contact: Biomechanical and pathological evidence from the Southeastern Borderlands. Paper presented, American Association of Physical Anthropologists, Miami.

Larsen CS, Schoeninger MJ, Hutchinson DL, Russell KF, and Ruff CB (1990) Beyond demographic collapse: Biological adaptation and change in native populations of *La Florida*. In DH Thomas (ed.): Columbian Consequences, vol. 2, Archaeological and Historical Perspectives on the

Spanish Borderlands East. Washington, DC: Smithsonian Institution Press, pp. 409–428.

Larsen CS, Ruff CB, Schoeninger MJ, and Hutchinson DL (1992) Population decline and extinction in *La Florida*. In JW Verano and DH Ubelaker (eds.): Disease and Demography in the Americas. Washington, DC: Smithsonian Institution Press, pp. 25–39.

Lester CW, and Shapiro HL (1968) Vertebral arch defects in the lumbar vertebrae of pre-historic American Eskimos. A study of skeletons in the American Museum of Natural History chiefly from Point Hope, Alaska. American Journal of Physical Anthropology *28*:43–48.

Merbs CF (1983) Patterns of Activity-Induced Pathology in a Canadian Inuit Population. Archaeological Survey of Canada Paper 119.

Merbs CF (1989) Spondylolysis: Its nature and anthropological significance. International Journal of Anthropology 4:163–169.

Nagurka ML, and Hayes WC (1980) An interactive graphics package for calculating cross-sectional properties of complex shapes. Journal of Biomechanics *13*:59–64.

Reitz EJ (1988) Evidence for coastal adaptations in Georgia and South Carolina. Archaeology of Eastern North America *16*:137–158.

Ruff CB (1984) Allometry between length and cross-sectional dimensions of the femur and tibia in *Homo sapiens sapiens*. American Journal of Physical Anthropology *65*:347–348.

Ruff CB (1987) Sexual dimorphism in human lower limb bone structure: Relationship to subsistence strategy and sexual division of labor. Journal of Human Evolution *16*:391–416.

Ruff CB (1992) Biomechanical analyses of archaeological human skeletal samples. In SR Saunders and MA Katzenberg (eds.): The Skeletal Biology of Past Peoples: Advances in Research Methods. New York: Wiley-Liss, pp. 37–58.

Ruff CB, and Hayes WC (1983) Cross-sectional geometry of Pecos Pueblo femora and tibiae—A biomechanical investigation: I. Method and general patterns of variation. American Journal of Physical Anthropology *60*:359–381.

Ruff CB, and Larsen CS (1990) Postcranial biomechanical adaptations to subsistence strategy changes on the Georgia coast. In CS Larsen (ed.): The Archaeology of Mission Santa Catalina de Guale: 2. Biocultural Interpretations of a Population in Transition. Anthropological Papers of the American Museum of Natural History *68*: 94–120.

Ruff CB, Larsen CS, and Hayes WC (1984) Structural changes in the femur with the transition to agriculture on the Georgia coast. American Journal of Physical Anthropology *64*:125–136.

Ruff CB, Scott WW, and Liu AY-C (1991) Articular and diaphyseal remodeling of the proximal femur with changes in body mass in adults. American Journal of Physical Anthropology *86*:397–413.

Saunders RA (1992) Continuity and Change in Guale Indian Pottery, A.D. 1350–1702. Ph.D. dissertation, Department of Anthropology, University of Florida, Gainesville.

Smith BD (1989) Origins of agriculture in eastern North America. Science *246*:1566–1571.

Stewart TD (1931) Incidence of separate neural arch in the lumbar vertebrae of Eskimos. American Journal of Physical Anthropology *16*:51–62.

Stewart TD (1979) Patterning of skeletal pathologies and epidemiology. In WS Laughlin and AB Harper (eds.): The First Americans: Origins, Affinities, and Adaptations. New York: Gustav Fischer, pp. 257–274.

Wiltse LL, Widell EH, and Jackson DW (1975) Fatigue fracture: The basic lesion in isthmic spondylolisthesis. Journal of Bone and Joint Surgery 57-A:17–22.

Chapter 4

Pilgrim's Progress and Praying Indians: The Biocultural Consequences of Contact in Southern New England

Brenda J. Baker

Department of Sociology and Anthropology, Moorhead State University, Moorhead, MN 56563

Epidemics of European diseases were a primary cause of depopulation among the indigenous inhabitants of southern New England during the seventeenth century (Cook, 1973, 1976). Equally devastating was the disruption of political-economic organization and traditional settlement and subsistence patterns that permitted English expansion throughout the region. In this chapter, the biocultural consequences of contact in southern New England are evaluated through documentary evidence, recent archaeological research, and analyses of human remains from seventeenth-century Native-American cemeteries. Although cemetery data and skeletal material from this period are sparse, much can be learned about the effects of contact on traditional customs, social organization, and health by examining extant information on burial practices, demography, and skeletal pathology.

EARLY EPIDEMICS AND ENGLISH COLONIZATION

There is no clear evidence that the sixteenth-century pandemics postulated by Dobyns (1983:313–325) reached the Northeast (see Snow and Lanphear, 1988:20–23). Early accounts, such as those of Martin Pring (1905) in 1603 and John Smith (1905: 223, 240–241) in 1614, describe robust, healthy inhabitants of the New England coast. Slightly later accounts, however, document the results of an epidemic that swept through previously populous coastal villages. In May of 1619, Thomas Dermer (1905:251) wrote:

> I passed alongst the Coast where I found some antient Plantations, not long since populous now vtterly void; in other places a remnant remaines, but not free of sicknesse. Their disease the Plague, for wee might perceiue the sores of some that had escaped, who described the spots of such as usually die.

The etiology of the epidemic, which raged from 1616 to at least 1619, has been debated extensively. Although it is most often identified as bubonic plague (Williams, 1909; Cook, 1973; Dobyns, 1983:322; Snow and Lanphear, 1988), the epidemic has also been attributed to chickenpox (Hoornbeck, 1976–1977), viral hepatitis (Spiess and Spiess,

In the Wake of Contact: Biological Responses to Conquest, pages 35–45

1987), or a combination of two or more diseases over a period of three to six years (Kelley, 1991:112). Whatever its cause, mortality was extreme among coastal groups from the eastern shore of Narragansett Bay in present-day Rhode Island northward into southern Maine. The Massachusett and Pokanoket (Wampanoag) groups between Massachusetts Bay and Cape Cod were hardest hit, with a population decline estimated between 75% and 90% (Cook, 1976:30–31; Salisbury, 1982:103–105). The epidemic, however, did not spread inland (cf. Dobyns, 1983:322), and some groups like the Narragansett escaped it entirely.

The Pilgrims, landing at Plymouth in 1620, thus found the land on which they settled nearly deserted. Early colonists noted bones of unburied victims littering former villages, most of which had been abandoned, and overgrown fields (Salisbury, 1982:103). Survivors of the epidemic were forced to amalgamate their villages and realign political alliances, while the strength of their unaffected neighbors, the Narragansett, increased commensurately (Salisbury, 1982:105–106). As a result, the Mayflower's 102 passengers were able to establish alliances with once-powerful sachems, such as Massasoit, and gain a foothold in New England (see Salisbury, 1982:175–176).

The English form of colonization departed radically from that of the French and Spanish. The Puritans' goal was to remove themselves from "spiritually polluted" forms of religion (Salisbury, 1982:171) and persecution by the Church of England. Ministers, deprived of their parishes, were easily induced to emigrate. The unit of settlement was the plantation, which required a tract of land, a political structure, and a congregational church (Haller, 1951). Seven towns were "planted" by the Massachusetts Bay Company in 1630, resulting in an explosion of the English population. Between 1630 and 1640, the number of colonists increased from about 2,200 to 17,800 (Haller, 1951: 15–16), overwhelming the remnant coastal Indians and coming into conflict with inland groups.

Coinciding with an increase in the immigration of families to New England was a well-documented epidemic of smallpox that struck all Northeastern Indian groups in 1633–1634 (see Cook, 1973; Duffy, 1953: 43–44; Salisbury, 1982:190–191; Snow and Lanphear, 1988). Based on documentary evidence, population estimates for the first decade of the seventeenth century (preepidemic) and mid-seventeenth century (postepidemic) indicate a reduction of 86% among the coastal Massachusett and Narragansett (from 44,000 to 6,400) to 95% among the Pocumtuck (18,400 to 920) of the Connecticut Valley (Snow and Lanphear, 1988:Table 1; cf. Jennings's [1975] estimates derived from Gookin's [1970] figures and Cook's [1976] substantially lower preepidemic population estimates but similar reduction rates). A reduction of 81% is estimated for the Mohegan and Pequot of the lower Connecticut River valley (Snow and Lanphear, 1988:Table 1). Beginning in 1632, however, these groups were also decimated by wars with the neighboring Narragansett and subsequently with the English colonists of Connecticut. By 1637, the Pequot were no longer a distinct group, and the Narragansett were isolated.

The earliest missionary effort was initiated by Thomas Mayhew, Jr., on Martha's Vineyard (Brenner, 1984:99–100). By the mid-seventeenth century, a missionary program was established in the Massachusetts Bay Colony by John Eliot, a Puritan preacher. Seven "Praying Towns" were created in eastern Massachusetts (Fig. 1) between 1651 and 1660 for the area's remaining indigenous people. Before King Philip's War (1675–1676), 14 Praying Towns had been established with an indigenous population of approximately 1,100 (Gookin, 1970:87). The stated purpose of these missions was to "civilize" the natives and convert them to

Christianity (Gookin, 1970:45–63; see also Axtell, 1985:133–136). Isolation, rather than assimilation, was an unstated goal of such efforts (Brenner, 1980:147–148; Axtell, 1985: 139; Carlson, 1986:1). Viewed as a buffer against hostile groups to the west, the English even supplied the Praying Indians with arms (Brenner, 1980:145; Axtell, 1985:148).

Native American groups within and beyond New England are often portrayed as hapless victims subjugated by colonists. Salisbury (1982:239), for example, states that "resisting this expansion was rendered virtually hopeless because of their severe epidemiological and, hence, demographic disadvantages and because of the colonists' readiness and ability to seize on those disadvantages." Such statements deny Native Americans any role; however, they did not merely react to circumstances over which they exercised little control, but made active choices. The Pequot, for example, invited the Massachusetts Bay Company to establish settlements in Connecticut in 1634, ostensibly to secure a truce with the Narragansett. Likewise, ethnohistorical records reveal that the inhabitants of Praying Towns followed traditional lines of political succession (Axtell, 1985:142–147), and they exercised political autonomy to the extent that some became allies of King Philip in his war against the colonists (Brenner, 1980).

Because historical documents were written from the perspective of the English (see Simmons, 1981), they do not reflect the Native American point of view. Thus, we must turn to the archaeological record to assess the degree to which the indigenous groups chose to accommodate or resist European encroachment in seventeenth-century New England.

ARCHAEOLOGICAL INVESTIGATIONS

Archaeological investigations of Native American contact period sites to date are sparse in New England. Many studies have involved reconnaissance and remote sensing to determine the location of Praying Towns or other habitation sites and associated cemeteries, along with limited test excavations (e.g., Carlson, 1986; Mulholland et al., 1986; Gumaer et al., 1987; Carlson et al., 1992). Excavations at four seventeenth-century cemetery sites in Massachusetts and Rhode Island (Fig. 1), however, have yielded considerable information on contact period burial practices and a significant body of biological data.

The earliest of these contact period sites is the West Ferry cemetery in Jamestown, Rhode Island, which contained 59 Narragansett graves dating between 1620 and 1660 (Simmons, 1970). Bodies were generally flexed and placed on their right sides in oval pits, the long axes of which shared a common orientation. The head was toward the southwest, the traditional location of Cautantowwit's house where souls continued in the afterlife (Simmons, 1970, 1978; Robinson et al., 1985). Grave goods of European and native manufacture at West Ferry were common; those of European derivation included items of brass, copper, iron, and pewter (Simmons, 1970).

Another Narragansett cemetery, RI1000 in North Kingstown, Rhode Island, contained 56 individuals (Robinson et al., 1985). Grave goods of European origin predominated, dating the cemetery from 1650 to 1670. RI1000, therefore, postdates both the 1633 smallpox epidemic and English settlement in the area. Like the earlier West Ferry cemetery, all burials were flexed with heads oriented to the southwest or south, and they were generally laid on their right sides (Robinson et al., 1985:114, 124). Graves were placed closer to each other than at West Ferry and were more rectangular in outline. Consistent burial orientation and posture at the two sites, however, indicate persistence of indigenous ritual and religious beliefs among the Narragansett throughout the seventeenth century (Robinson et al., 1985).

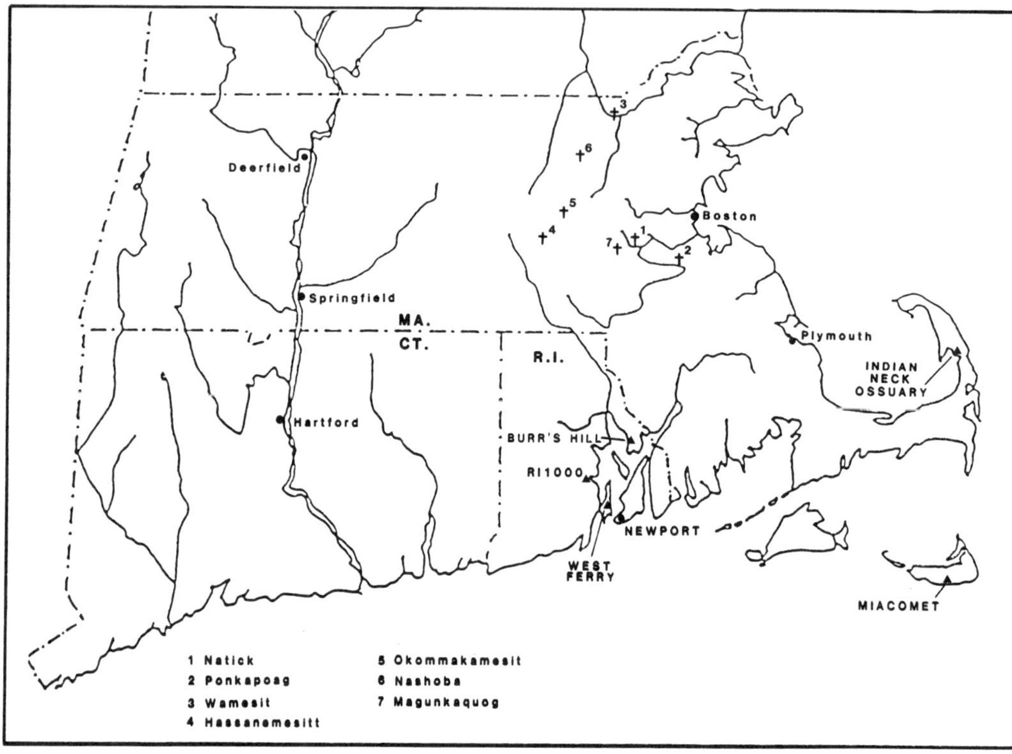

Fig. 1. Location of the original Praying Indian Towns and other significant sites in southern New England.

The Burr's Hill site, a Wampanoag burial ground in Warren, Rhode Island, was also used during the third quarter of the seventeenth century (most artifacts date between 1655 and 1677 [Gibson, 1980:22]). Forty-two graves were excavated in 1913 when gravel quarrying disturbed human remains. Gibson's (1980) burial inventory, reconstructed from the original field notes and museum records, includes information on burial form. Most skeletons were flexed and laid on their right sides with heads oriented to the southwest, south, or southeast in the traditional manner (Gibson, 1980:13). Many were wrapped in matting or blankets, and some had been covered with bark boards or contained traces of red ocher.

At least two individuals at Burr's Hill were buried with their legs extended in the European fashion. Because these burials were oriented in the same direction as the flexed interments and contained a full complement of grave goods, Gibson (1980:13) concludes that the body posture was "incorporated into the traditional mortuary complex, in much the same manner that European trade wares were." At Burr's Hill, 86% (36/42) of the burials included grave goods, compared to only 50% (24/48) of previously undisturbed burials at West Ferry, with the former site yielding three times the number of artifacts as the latter (Gibson, 1980:14). The relative wealth at Burr's Hill is attributed by Gibson (1980:14) to its position in or near the principal village of the chief sachem (Massasoit, who was followed by his son, Philip) and, hence, this group's greater access to trade goods. It is also possible that the proliferation of Euramerican goods (e.g., beads, bottles, pipes, iron hoes, gun parts) in the Burr's Hill graves represents a greater degree of acculturation among the Wampanoag (cf. Brenner, 1988:151). Gibson (1980:23), however, cautions that the proportion of European to aboriginal objects in the graves may not reflect their ratios in daily use. European objects may, instead, have been preferred as grave offerings because of their novelty and prestige value, and as symbols of power and authority (Gibson, 1980:23; Brenner, 1988:172–176; see also Axtell, 1981:110–128; Crosby, 1988: 187–189, 193).

At least 10 of the burials at Burr's Hill were multiple interments. Like the solitary multiple interment at West Ferry, four of these graves contained one adult and one child (Gibson, 1980:13). One grave at Burr's Hill, however, contained seven individuals, including one child and one infant or fetus. Another grave contained six complete skeletons, along with two skulls in a large tub and a kettle full of disarticulated ribs and other small bones. The excavators postulated that these people were "victims of a scourge" (quoted in Gibson, 1980:13). The single case in which pathological lesions were reported at Burr's Hill lends some credence to this possibility. Soft tissue on the left side and top of the skull of one individual, preserved by a large kettle over the head and shoulders, revealed small pustules (Gibson, 1980:17). Gibson (1980:24) suggests that "the high incidence of multiple interments at Burr's Hill might have been due to a smallpox epidemic." This particular individual, however, was buried alone. In addition, secondary interment is inconsistent with the assumption that mass graves resulted from hasty burial of people who died in an epidemic.

Recent excavations at a cemetery associated with the second Praying Town of Ponkapoag, established in 1652 in what is now Canton, Massachusetts, reveal mortuary practices that differed from those at the other seventeenth-century cemeteries. The unmarked site was partially bulldozed in 1988, and parts of 12 individuals were found in disturbed contexts (Simon, 1990). Another 32 burials were observed at the site, but only six were completely excavated. Artifacts accompanying the burials include kaolin pipes, metal pots, and glass or shell beads, which indicate that the cemetery was used between 1652 and 1720 (Simon, 1990). Of the 32 graves, 31 have rectangular grave shafts or coffin remains suggestive of Christian burial practices (Simon, 1990). Eight of the 10 skeletons observed in situ were supine and extended.

Despite these differences, 58% of the Ponkapoag graves were oriented on a southwest-northeast axis and, where skeletal remains were observable, the skull was at the southwestern, southern, or western end of the grave (Simon, 1990). The preference for traditional orientation and continuation of the practice of placing offerings in graves indicate some persistence of indigenous customs (see also Axtell, 1981:110–128). In comparison to the Narragansett and Wampanoag cemeteries, however, it appears that compromises were made by the inhabitants of Ponkapoag.

SKELETAL BIOLOGY

Demography

Demographic data of varying quality can be gleaned from the skeletal samples of the seventeenth-century cemeteries described above, while information on health status is available for RI1000 and Ponkapoag. Several limitations inhibit full study of these series. Burr's Hill was excavated by an amateur archaeologist in the early twentieth century, following years of disturbance to the site (Gibson, 1980:9–10). Little attempt was made to determine the age or sex of the burials, and pathological lesions were noted in only one case. The skeletal remains from Burr's Hill were not retained for further study (Gibson, 1980:14). The West Ferry remains, excavated in the 1960s, were rebur-

TABLE 1. Age and Sex Distribution of Burials from Postcontact Sites in New England[a]

	West Ferry				RI1000				Ponkapoag			
Age	M	F	U	T	M	F	U	T	M	F	U	T
0–2.5			4	4			0	0			2	2
2.5–3			3	3			1	1			0	0
3–5			7	7			9	9			0	0
5–10			5	5			3	3			0	0
10–15	0	1	0	1	1	4	4	9	1	0	0	1
15–20	0	2	0	2	2	7	0	9	0	1	0	1
20–25	0	0	0	0	1	1	0	2	0	2	0	2
25–30	0	1	0	1	3	2	0	5	1	2	0	3
30–35	3	2	0	5	1	1	0	2	0	1	0	1
35–40	2	1	0	3	2	2	0	4	2	3	0	5
40–45	2	1	0	3	2	5	0	7	0	0	0	0
45–50	1	0	0	1	1	1	0	2	0	0	0	0
50+	8	6	0	14	0	3	0	3	0	0	0	0
Unknown	2	1	1	4	0	0	0	0	1	2	0	3
Total	18	15	20	53	13	26	17	56	5	11	2	18
M:F ratio	1.2:1.0				0.5:1.0				0.45:1.0			

[a]Age is in years. M, male; F, female; U, unknown; T, total.
(West Ferry data derived from Simmons, 1970; RI1000 data from Kelley et al., 1987b:4; Ponkapoag data from Simon, 1990.)

ied in the 1970s (Kelley et al., 1987b:3, 16–17) without any analysis by a biological anthropologist. The Ponkapoag cemetery has not been entirely excavated, thus the sample size is small (Simon, 1990). In contrast, the RI1000 cemetery was completely excavated in the early 1980s, and the skeletal remains were analyzed by biological anthropologists. Despite these limitations, the information obtained from the skeletal remains provides an important, though often overlooked, perspective on Native American response to European contact.

Comparison of the age and sex distribution of individuals interred at the West Ferry, RI1000, and Ponkapoag cemeteries reveals substantial differences (Table 1). Most striking is the male to female sex ratio. At the earliest site, West Ferry, the sex ratio is nearly equal. At the later two sites, however, women outnumber men by a ratio of 2:1. The paucity of men at RI1000 and Ponkapoag may be due to the participation of men in conflicts with other groups (see Axtell, 1981:262–263), removal of young males to boarding schools (see Axtell, 1985:179–217), or the employment of men as laborers outside their own communities (Simon, 1990). Additionally, in the Praying Towns, male neophytes frequently ran away (see Brenner, 1980:145–147). All such activities could result in the burial of many men elsewhere. These factors were all common in areas outside of New England as well (see Baker and Kealhofer, n.d.).

In contrast, men outnumber women by nearly 2:1 at Burr's Hill (10 females and 22 males out of 74 individuals). Gender at Burr's Hill, however, is inferred solely from associated grave goods. (Gibson [1980:14; see also Simmons, 1970:44–47, and Crosby, 1988:186] indicates that pipes, gun parts, and fishing equipment are found with men, while hoes and pestles are found with women.) An additional 32 individuals (half of the apparent adult age group) cannot be categorized on this basis. The inferred sex ratio at Burr's Hill is, thus, tenuous. Gibson (1980:16) found that graves of men tended to contain more offerings than those of women. Based on this observation, the majority of indeterminate-sex individuals were probably female. The resulting sex ratio could, therefore, be similar to RI1000 and Ponkapoag, with women far outnumbering men.

The age distributions at RI1000 and Ponkapoag indicate additional deviations from what would be considered a realistic mortality profile. Individuals under age 2.5 are absent at RI1000, in contrast to the other sites (Table 1; at Burr's Hill, five "babies" were recorded). Kelley and coworkers (1987b:3; Robinson et al., 1985:114) suggest that members of this age group were intentionally excluded from burial in the RI1000 cemetery. They extend this observation to West Ferry (Kelley et al., 1987b:3), where age was typically reported in lengthy spans. In two cases, however, age at West Ferry was specified as either less than one year or approximately one year (Simmons, 1970), thus refuting the contention that infants were excluded from this cemetery.

At West Ferry, children under the age of 10 years comprise 35.8% of the sample (Table 1). At RI1000, children comprise 23.2% of the skeletal series, and 11.1% at Ponkapoag. Children form a small percentage (nine of 57 or 15.8%) of the Burr's Hill interments for which the adult or child category was specified. Adolescents (10–20 years of age) account for only 5.7% of the burials at West Ferry and 11.1% at Ponkapoag (only one individual at Burr's Hill can be assigned to this group). In contrast, 32.1% (18 of 56) of the RI1000 burials are adolescents. Adults form 49% of the West Ferry burials, 39.3% at RI1000, and 61.1% at Ponkapoag. At West Ferry, more than half of the adults are over age 50. The adults at the other two sites are less than 40 years of age at death.

What might account for these disparate demographic profiles? One might infer that the high mortality among small children and the elderly seen at the West Ferry site represents a fairly "normal," or precontact, pattern. Subadults, however, may be underrepresented at the other sites due to burial elsewhere (RI1000), unsophisticated excavation techniques (Burr's Hill), or incomplete excavation (Ponkapoag). The preponderance of older adults at West Ferry may reflect a reduced birth rate after contact (cf. Larsen et al., 1990:417–418). Age estimates for the West Ferry remains, however, were not made by a biological anthropologist, and the number of elderly adults reported by Simmons (1970) could, therefore, be overestimated.

Unfortunately, little comparative data exist from the Late Woodland period in southern New England. Most Late Woodland interments were isolated or occurred in small groups rather than within large cemeteries (Brenner, 1988:154–156). The Indian Neck ossuary on Cape Cod (Fig. 1) is an exception. Radiocarbon-dated between A.D. 935 and 1020 (McManamon and Bradley, 1986:18), this ossuary is isolated both temporally and spatially (Magennis, 1986:51). The fragmentary and commingled nature of the remains at this site renders direct comparison with later accretional cemeteries difficult. At least 25 adults and 22 subadults were found in the unburned portion of the ossuary. In addition, at least six adults and three children were cremated (Magennis, 1986:57–59). Magennis (1986:68–72) identified 12 subadults in the 0–5 year age group, three in the 5–10 year group, two in the 10–15 year category, and three between ages 15 and 20 years. Thus, children under age 10 form at least 26.8% of the minimum number of individuals, whereas adolescents comprise only 8.9% of the ossuary sample. Adults account for 55.4% of the sample. Based on mandibular dental attrition, five individuals (16.1% of all adults) were over age 50 (Magennis, 1986:Table 1.9). Gender determinations, where possible, indicate that males and females are nearly equally represented (Magennis, 1986:65–67).

The demographic profile from the Indian Neck ossuary is most like West Ferry in its relative proportions of children, adolescents, and adults, as well as males and females. Those over age 50, however, form a much lower percentage of the adult sample from the ossuary. The West Ferry cemetery was used between 1620 and 1660, when the effects of European contact were clearly felt in the area. Because some of the individuals interred in this cemetery may have been victims of the 1633 smallpox epidemic, the mortality profile cannot be assumed to reflect a pattern typical of the period before contact.

Pathology

The large proportion of adolescents buried at RI1000 (Table 1) is exceptional. Kelley and colleagues (1987b; Robinson et al., 1985:116–119) attribute the high rate of adolescent mortality at RI1000 to the presence of widespread tuberculosis during the seventeenth century. They assert that 17 of 56 individuals (30%) exhibit lesions of the spine, rib, and/or hip that are indicative of tuberculosis. Eight skeletons display smooth-walled resorptive lesions on the bodies of the lower thoracic and lumbar vertebrae. Twelve individuals, including four with corresponding vertebral involvement, present periostitic lesions on the internal surfaces of ribs. Three individuals, two of whom also have rib or vertebral lesions, display smooth-walled, oval lesions on femoral heads.

It should be cautioned that rib lesions like those found in the RI1000 skeletons may not be pathognomonic (cf. Kelley, 1991:118); these lesions are also found in association with other respiratory infections, such as pneumonia or mycoses (Roberts et al., 1992). Kelley (1991:115) recognizes documentary evidence of other infectious diseases, such as pneumonia, among the Narragansett. A more conservative estimate for the frequency of skeletal tuberculosis would exclude those with rib lesions alone, leaving 10 cases or about 18% of the RI1000 group with skeletal manifestations of the disease. This is still a high rate of skeletal infection (cf. Steinbock, 1976:175).

It might be anticipated that the individuals buried at the Ponkapoag cemetery would show more tuberculous lesions than at the Narragansett cemetery due to the further disruption of precontact lifeways and the concentration of people in the Praying Town. This does not appear to be the case. Of the 18 individuals represented at Ponkapoag, only two (11.1%) exhibit lesions indicative of possible infection, neither of which is diagnostic of tuberculosis. Evidence of infection consists of mild periostitis on the left tibia of a male aged 35–45 years and a rib lesion identified as a possible osteoma or calcification from a pulmonary infection in another male of the same age (Simon, 1990:Table 2).

Evidence of trauma at RI1000 is restricted to three examples (7.5%) of Colles' fractures and three cases (7.5%) of vertebral compression fractures (Kelley et al., 1987b:9). At Ponkapoag, two cases (11.1%) of vertebral trauma are reported (Simon, 1990), which may have been caused by blows to the spine. The trauma at Ponkapoag is suggestive of violence, whereas at RI1000 lesions are attributable to falls and osteoporosis.

Osteoarthritis is frequent in both groups, affecting 60% of females and 36% of males at RI1000 (Kelley et al., 1987b:13). At Ponkapoag, 36% of females and 40% of males are affected (Simon, 1990:Table 2).

Dental caries are present in 41 of 53 individuals (77.4%) at RI1000 (Kelley et al., 1987a:115). At Ponkapoag, only two of 18 individuals (11.1%) have carious teeth (Simon, 1990). The high rate of caries (32.9%) at RI1000 may be due to an increase in the consumption of maize, flour, or sugar (Kelley et al., 1987a:121–122). Again, it might be expected that dental caries would be more common at the later Praying Town cemetery where Native Americans were supposed to live a more European lifestyle. This is not the case, however. This observation lends support to the argument that Native Americans at the Praying Towns had considerable autonomy (Brenner, 1980, 1984) and may have continued traditional subsistence patterns.

CONCLUSION

The ethnohistorical records describe once-populous native groups that were ravaged by epidemics in 1616–1619 and 1633, as well as by warfare. Despite evidence for the persistence of traditional customs, the demographic profiles of four seventeenth-century cemeteries in southern New England reveal disparities in their sex ratios and age profiles. The preponderance of females at the later RI1000 and Ponkapoag cemeteries indicates that males were likely removed from the community, possibly due to engagement in warfare or wage labor.

The later Narragansett cemetery, RI1000, does not exhibit a realistic mortality profile. The large proportion of individuals exhibiting evidence of infection cannot be explained simply as a consequence of epidemics. The frequency of skeletal involvement and its association with chronic conditions, including tuberculosis, suggests that these people were experiencing diseases that they survived for some time. Many people at RI1000 experienced infections affecting bone, but this pattern does not appear to extend to the Praying Towns of eastern Massachusetts. Given the lack of older adults among those interred at the Ponkapoag cemetery, it is possible that those settled in mission towns were unable to survive long enough for diseases like tuberculosis to affect the skeleton.

The paucity of Native American skeletal samples spanning the period prior to the arrival of Europeans through the eighteenth century in southern New England makes it difficult to document biological changes resulting from increased interaction. Nonetheless, extant data from four seventeenth-century Native American cemeteries indicate a persistence of precontact mortuary customs despite considerable alterations in demographic structure and health status through time.

ACKNOWLEDGMENTS

Brona Simon, State Archaeologist, Massachusetts Historical Commission, shared her unpublished work on the Ponkapoag cemetery with me. Helpful comments on a draft of this chapter were provided by Claire Carlson and Dena Dincauze of the University of Massachusetts, Amherst, and by the volume's editors.

REFERENCES CITED

Axtell J (1981) The European and the Indian: Essays in the Ethnohistory of Colonial North America. New York: Oxford University Press.

Axtell J (1985) The Invasion Within: The Contest of Cultures in Colonial North America. New York: Oxford University Press.

Baker BJ, and Kealhofer L (eds.) (n.d.) Disease and Biocultural Frontiers: Native American Adaptation in the Spanish Borderlands. Submitted for publication.

Brenner EM (1980) To pray or be prey: That is the question. Strategies for cultural autonomy of Massachusetts Praying Town Indians. Ethnohistory 27:135–152.

Brenner EM (1984) Strategies for Autonomy: An Analysis of Ethnic Mobilization in Seventeenth Century Southern New England. Ph.D. dissertation, Department of Anthropology, University of Massachusetts, Amherst.

Brenner EM (1988) Sociopolitical implications of mortuary ritual remains in 17th-century native southern New England. In MP Leone and PB Potter, Jr. (eds.): The Recovery of Meaning. Washington, DC: Smithsonian Institution Press, pp. 147–181.

Carlson CC (1986) Archival and Archaeological Research Report on the Configuration of the Seven Original 17th Century Praying Indian Towns of the Massachusetts Bay Colony. University of Massachusetts Archaeological Services Report 22.

Carlson CC, Little EA, Gumaer DR, Loparto LW, and Baker BJ (1992) Archaeological Survey and Historical Background Research for the Miacomet

Indian Village and Burial Ground, Nantucket, Massachusetts. University of Massachusetts Archaeological Services Report 88.

Cook SF (1973) The significance of disease in the extinction of the New England Indians. Human Biology 45:485–508.

Cook SF (1976) The Indian Population of New England in the Seventeenth Century. Berkeley: University of California Press.

Crosby CA (1988) From myth to history, or why King Philip's ghost walks abroad. In MP Leone and PB Potter, Jr. (eds.): The Recovery of Meaning. Washington, DC: Smithsonian Institution Press, pp. 183–209.

Dermer T (1905) To his worshipfull friend M. Samvel Pvrchas, preacher of the Word, at the Church a little within Ludgate, London. Originally published in 1625. In GP Winship (ed.): Sailors Narratives of Voyages along the New England Coast 1524–1624. Boston: Houghton, Mifflin & Company, pp. 251–258.

Dobyns HF (1983) Their Number Become Thinned. Knoxville: University of Tennessee Press.

Duffy J (1953) Epidemics in Colonial America. Baton Rouge: Louisiana State University Press.

Gibson SG (1980) Introduction. In SG Gibson (ed.): Burr's Hill: A 17th Century Burial Ground in Warren, Rhode Island. Providence: Haffenreffer Museum of Anthropology, Brown University, pp. 9–24.

Gookin D (1970) Historical Collections of the Indians of New England. Towtaid, New Jersey. Originally published 1792, Massachusetts Historical Society Collections.

Gumaer DR, Johnson R, and McArdle AH (1987) Archaeological Locational and Remote Sensing Survey of an Unmarked Historic Cemetary [sic] and Vicinity, Mashpee, Massachusetts. University of Massachusetts Archaeological Services Report 48.

Haller W Jr. (1951) The Puritan Frontier: Town-Planting in New England Colonial Development, 1630–1660. New York: Columbia University Press.

Hoornbeck B (1976–1977) An investigation into the cause or causes of the epidemic which decimated the Indian population of New England 1616–1619. New Hampshire Archaeologist 19: 35–46.

Jennings F (1975) The Invasion of America. Chapel Hill: University of North Carolina Press.

Kelley MA (1991) Ethnohistorical accounts as a method of assessing health, disease, and population decline among Native Americans. In DJ Ortner and AC Aufderheide (eds.): Human Paleopathology: Current Syntheses and Future Options. Washington, DC: Smithsonian Institution Press, pp. 111–118.

Kelley MA, Barrett TG, and Saunders SD (1987a) Diet, dental disease, and transition in Northeastern Native Americans. Man in the Northeast 33:113–125.

Kelley MA, Sledzik PS, and Murphy SP (1987b) Health, demographics, and physical constitution in seventeenth century Rhode Island Indians. Man in the Northeast 34:1–25.

Larsen CS, Schoeninger MJ, Hutchinson DL, Russell KF, and Ruff CB (1990) Beyond demographic collapse: Biological adaptation and change in native populations of La Florida. In DH Thomas (ed.): Columbian Consequences, vol. 2, Archaeological and Historical Perspectives on the Spanish Borderlands East. Washington, DC: Smithsonian Institution Press, pp. 409–428.

Magennis AL (1986) Part II: The physical anthropology of the Indian Neck ossuary. In The Indian Neck Ossuary: Chapters in the Archeology of Cape Cod, V. Cultural Resources Management Study No. 17, Division of Cultural Resources, North Atlantic Regional Office, National Park Service, pp. 49–183.

McManamon FP, and Bradley JW (1986) Part I: The Indian Neck ossuary and Late Woodland prehistory in southern New England. In The Indian Neck Ossuary: Chapters in the Archeology of Cape Cod, V. Cultural Resources Management Study No. 17, Division of Cultural Resources, North Atlantic Regional Office, National Park Service, pp. 1–47.

Mulholland MT, Savulis ER, and Gumaer DR (1986) Archaeological Remote Sensing and Historical Research Survey for the Yeadon Parcel at the Hassanomisco Indian Burying Ground, Grafton, Massachusetts. University of Massachusetts Archaeological Services Report 38.

Pring M (1905) A voyage set out from the citie of Bristoll at the charge of the chiefest merchants and inhabitants of the said citie with a small ship and a barke for the discouerie of the north part of Virginia. Originally published in 1625. In GP Winship (ed.): Sailors Narratives of Voyages along the New England Coast 1524–1624. Boston: Houghton, Mifflin & Company, pp. 53–63.

Roberts C, Manchester K, and Lucy D (1992) Rib lesions in palaeopathology: A modern study and its relevance. Paper presented, Paleopathology Association, Las Vegas.

Robinson PA, Kelley MA, and Rubertone PE (1985) Preliminary biocultural interpretations from a seventeenth century Narragansett Indian cemetery in Rhode Island. In W Fitzhugh (ed.): Cultures in Contact: The Impact of European Contacts on Native American Cultural Institutions, A.D. 1000–1800. Washington, DC: Smithsonian Institution Press, pp. 107–130.

Salisbury N (1982) Manitou and Providence: Indians, Europeans, and the Making of New England, 1500–1643. New York: Oxford University Press.

Simmons WS (1970) Cautantowwit's House. Providence: Brown University Press.

Simmons WS (1978) Narragansett. In BG Trigger (ed.): Northeast. Handbook of North American Indians, vol. 15. Washington, DC: Smithsonian Institution, pp. 190–197.

Simmons WS (1981) Cultural bias in the New England Puritans' perception of Indians. William and Mary Quarterly *38*:56–72.

Simon BG (1990) Native American culture change and persistence in Contact period New England: Analysis of mortuary data from a Praying Indian burial ground in Massachusetts. Paper presented, Society for American Archaeology, Las Vegas.

Smith J (1905) A description of New England. Originally published in 1616. In GP Winship (ed.): Sailors Narratives of Voyages along the New England Coast 1524–1624. Boston: Houghton, Mifflin & Company, pp. 213–247.

Snow DR, and Lanphear KM (1988) European contact and Indian depopulation in the Northeast: The timing of the first epidemics. Ethnohistory *35*:15–33.

Spiess AE, and Spiess BD (1987) New England pandemic of 1616–1622: Cause and archaeological implication. Man in the Northeast *34*:71–83.

Steinbock RT (1976) Paleopathological Diagnosis and Interpretation. Springfield: Charles C Thomas.

Williams HU (1909) The epidemic of the Indians of New England, 1616–1620, with remarks on Native American infections. Johns Hopkins Hospital Bulletin *20*:340–349.

Chapter 5

Evidence from Ossuaries: The Effect of Contact on the Health of Iroquoians

Susan Pfeiffer and Scott I. Fairgrieve

School of Human Biology, University of Guelph, Guelph, Ontario N1G 2W1, Canada (S.P.); and Departments of Sociology and Anthropology and Biology, Laurentian University, Sudbury, Ontario P3E 2C6, Canada (S.I.F.)

When European explorers, traders, and priests contacted the Iroquoian speakers of Upper Canada (southern Ontario) in the early seventeenth century, they encountered a complex tribal society based on horticulture, hunting, and fishing. These first Europeans included numerous observers whose reports to France have been the subject of much scholarly interest (e.g., Tooker, 1964; Trigger, 1969, 1976, 1985; Heidenreich, 1971). Several reliable observers, including Champlain (Biggar, 1929:160–163), Sagard (Wrong, 1939:211), and Brebeuf (Thwaites, 1896–1901, 10:279–305; Kidd, 1953:372–375) described aspects of the Feast of the Dead and the construction of large, single-pit cemeteries known as ossuaries. They suggested that virtually all deceased members of the group were traditionally buried together at regular intervals. Such practices should allow us to reconstruct the groups' demographic and epidemiological characteristics with some accuracy, although the inherent limitations of ossuary samples create certain challenges to such an analysis.

Firsthand observations of the Huron have influenced interpretations of life in both contact and late precontact times. For example, Brebeuf referred to the Huron as being "more healthy than we" (Thwaites, 1896–1901, 38:257). Jacques Cartier, who visited the Iroquoian village of Hochelaga in 1535, credited a native tonic made from hemlock or pine with curing his men of scurvy within a week (Fagan, 1991). While such observations can be tested with osteological evidence, they also influence our assumptions, so that it is difficult to approach skeletal data without bias.

By 1650, decimated by warfare and epidemics, the last 500 Huron abandoned their homeland, ending hundreds of years of successful adaptation in Ontario. We know that smallpox, measles, and influenza had a devastating impact on the Huron and their neighbors from 1634 through 1640 (Trigger, 1976). To what extent were these epidemics a new phenomenon among Iroquoian villages? Did new European diseases disturb a robust and healthy population, or did demographic decline actually precede European contact? Can we generalize from ethnographic and archaeological knowledge of the Huron to deduce the fate of their neighbors?

Recent debate in the ethnohistorical and archaeological literature furnishes various viewpoints to explore. Warrick (1990) pro-

In the Wake of Contact: Biological Responses to Conquest, pages 47–61

poses that during the fourteenth century the maize farmers of south-central Ontario experienced a population explosion, with numbers growing from 8,000 to 30,000 between A.D. 1300 and A.D. 1430. This growth ceased prior to contact with Europeans and numbers remained stable during the sixteenth and early seventeenth centuries. Epidemics of Old World diseases did not affect the Huron-Petun until 1634, from which time the population loss was catastrophic.

Saunders and coworkers (1992) argue that the Ontario Iroquoians were probably already dealing with a heightened infection load well before European contact. The precontact record from the Balsam Lake region is considered exemplary of broader patterns showing greater population size, larger and more numerous villages, and increasingly poor hygiene within the villages. All of these features are consistent with a greater disease load.

In this chapter, we explore the osteological evidence from southern Ontario, specifically the role of diseases and ill health. Although the epidemics of 1634 and later are well documented, skeletal evidence must be examined for evidence of disruption, particularly prior to this time. By necessity, any exploration of change over time must also investigate the health characteristics of the precontact population. Within each of the following sections, we review the evidence from ossuaries regarding the state of health prior to contact as compared to after first potential contact (A.D. 1535). We include information not only from the Huron region but also from their neighbors, historically known as the Neutrals and the Petun.

THE NATURE OF IROQUOIAN OSSUARIES

Collective ossuary burial is a characteristic of the terminal Woodland period (beginning about A.D. 1000) (Wright, 1966). The size and characteristics of these ossuaries vary temporally and geographically, generally becoming larger. By the time of European contact, Huron ossuaries were quite large, often including the remains of over 500 people. Ethnographic and archaeological evidence indicates that the remains were thoroughly mixed and tightly packed during the Feast of the Dead (Pfeiffer, 1980a), but there are inconsistencies between the written and physical evidence (Sutton, 1988). For example, Europeans noted the inclusion of many grave goods, and late ossuaries include many iron axe heads, copper kettles, and beads. Precontact ossuaries are notable for their lack of preserved grave goods.

Much debate has centered on the European observation that newborn infants were not included in the ossuaries. "If a young baby, less than a month or two old, died, it was buried by the road, in order that it might enter a woman's womb and be reborn" (JR 10:273, cited in Tooker, 1964:132). In fact, the proportion of newborns in ossuaries is highly variable: some include large numbers of infants (e.g., Uxbridge), while others are apparently deficient (e.g., Kleinburg) (see Table 1). The number of infant burials in occupation sites also varies (Saunders and Spence, 1986). Such cultural practices may have been variable through space and time.

METHODOLOGICAL CONSIDERATIONS

The ideal reconstruction of protohistoric change would incorporate evidence from both paleopathology and paleodemography. The paleodemographic literature has explored methods and taphonomic considerations to establish valid age-at-death data sets (Katzenberg and White, 1979; Pfeiffer, 1985, 1986). Researchers have attempted to extract meaningful information by focusing on the age categories thought to be least problematical, such as the ratio of children (excluding infants) to adults (Jackes, 1986) and the ratio of young adults to others (Larocque, 1991). The one consistent observation that can be drawn from this work is

TABLE 1. Percentage of Different Age Groups of Subadults from Ontario Sites after ca. A.D. 1300 (Modified from Larocque 1991)

Site	< 20 years (%)[a]	15–20 years (%)[b]	< 3 years (%)[b]	References
Tabor Hill	49	76[c]	—	Churcher and Kenyon, 1960
Fairty	43	8[d]	35	Larocque, 1991
Glen Williams	40	14	18	Hartney, 1978
Garland	42	27[e]	—	Webb, 1969
Roebuck	49	10	39	Knowles, 1937
Uxbridge	32	9	53	Pfeiffer, 1983
Kleinburg	28	15[d]	17	Larocque, 1991
Milton	37	10	22	Hartney, 1978
Warminster	35	—	—	Harris, n.d.
Carton	21	42[f]	—	Hartney, 1978
Grimsby	43	26	—	Jackes, 1986
Ossossané	25	16	23	Larocque, 1991
Maurice	26	32[f]	18[g]	Jerkic, 1975

[a]Percentage of total individuals.
[b]Percentage of total subadults.
[c]16–20 years.
[d]After splitting the 12–18 year group into two groups of equal content.
[e]17–20 years.
[f]12–18 years.
[g]For the group 0–2 years.

that there is intersite variability, presumably a product of true demographic differences combined with cultural differences in ossuary construction, conditions of postdepositional preservation, and the application of different osteological techniques by separate osteologists. Overall, data are consistent with high infant and juvenile mortality and low life expectancy at birth, but they cannot be finessed into a more sensitive discriminator of short-term changes. At this time, the paleodemographic approach offers us no solid insights into health and well-being during the contact period.

The best paleopathological information will come from ossuaries that are structurally intact, fully excavated, well dated, and culturally linked to specific villages. No such sites exist, although some come closer to fulfilling these criteria than others. Because of a long-standing voluntary moratorium on the excavation of human skeletal remains in this region for research (as distinct from occasional salvage work), the most scientifically valuable sites were excavated at least 20 years ago, and much of the fundamental research is correspondingly old. Our discussion focuses on well-documented, relatively complete collections. The Ossossané ossuary, excavated in 1953, is thought to represent the remains of the Feast of the Dead observed by Jean de Brebeuf in 1636. Thus, it is very valuable, despite its rather poor physical preservation. Several sample characteristics are summarized in Table 2.

The physical mixing of elements within an ossuary necessitates an approach in which data must be collected on a bone-by-bone basis. However, most single elements do not yield accurate indicators of sex and age-at-death, so data often must be interpreted without individual characteristics. There is also strong evidence of taphonomic factors affecting preservation (Pfeiffer, 1986). Further, it is obviously not possible to examine the relationship of characteristics be-

TABLE 2. Summary Information for Southern Ontario Ossuaries

Site	Date	Minimum number of individuals	Map code	Major references
Tabor Hill	1300–1350	523	1	Churcher and Kenyon, 1960 Hartney, 1978 Kapches, 1979 Molto, 1983
Middleport	1400		2	Knowles, 1937 Wintemberg, 1948 Wright, 1966 Molto, 1983
Fairty	1400–1450	512	3	Anderson, 1963 Wright, 1966 Stothers and Metress, 1975 Molto, 1983 Jackes, 1986 Larocque, 1991
Syers	1300–1500	300	4	AARO, 1894–1895 Ramsden, 1977 Webb, 1972 Molto, 1983
Glen Williams	1400–1500	309	5	Hartney, 1978 Molto, 1983 Pfeiffer, 1991
Roebuck	1450–1530	84	6	Knowles, 1937 Anderson, 1972 Pendergast, 1975 Ramsden, 1977 Molto, 1983
Possible contact from this point				
Uxbridge	1490 ± 80	457	7	Cook, n.d. Molto, 1983 Pfeiffer, 1986, 1991 Gagné, 1990
Garland	1500–1550	198	8	Wright, 1966 Webb, 1969 Molto, 1983
Keffer	1500–1575	Incomplete excavation 50 crania examined	9	AARO, 1889–1890, 1907 Webb, 1972 Ramsden, 1977 Molto, 1983
Postcontact from this point				
Orchid	1500–1700[a]	519	10	Cybulski, 1966 White, 1966 Melbye, 1967, n.d. Clabeaux, 1967, 1977 Sirianni, 1967 Hartney, 1978 Molto, 1983 Stothers, 1977 Williamson, 1990 Birx, 1991

TABLE 2. Summary Information for Southern Ontario Ossuaries *(Continued)*

Site	Date	Minimum number of individuals	Map code	Major references
Carton	1585–1615	303	11	Sundick, n.d. Winnicki, n.d. Wright, 1974 Molto, 1983
Kleinburg	1585–1615	561	12	Pfeiffer, 1980a,b, 1985 Jackes, 1977 Pfeiffer and King, 1983 Patterson, 1984 Saunders and Melbye, 1990 Larocque, 1991 Molto, 1983
Ossossané	1636	419	13	Ridley, 1947 Kidd, 1953, 1954 Gruber, 1958 Heidenreich, 1971 Trigger, 1976, 1985 Ramsden, 1977 Katzenberg and White, 1979 Molto, 1983 Jackes, 1986 Saunders and Melbye, 1990 Larocque, 1991
Maurice	1640	132	14	Jerkic, 1975 Molto, 1983
Grimsby	1640–1650	374	15	Kenyon, 1977, n.d. Jackes, n.d.
Houghton (BeGw-2)	Historic	Unknown	16	Archaeological Services, 1990
Milton	Historic	109	17	Hartney, 1978
Gaetan	Historic	Not applicable Incomplete excavation	18	Melbye et al., n.d.
Warminster	Historic	250 (originally)[b]	19	McIlwraith, 1946, 1947 Harris, 1949, n.d. Mullen, 1990

[a]New uncorrected radiometric date, see Birx (1991) for discussion.
[b]Degradation of sample over time.

tween the various elements of the skeleton. This is a serious drawback for differential diagnosis. These shortcomings are balanced by some obvious strengths. Ossuaries include the remains of a culturally and temporally defined cohort with considerable genetic commonality (Pfeiffer et al., 1985) because they include most members of a community who died during a given period (roughly one decade). All bones and teeth were interred in a small space, thus in principle minimizing differential preservation.

PALEODIETARY INFORMATION

The introduction of maize into southern Ontario has been a focus for paleodietary research, and this work also offers insights

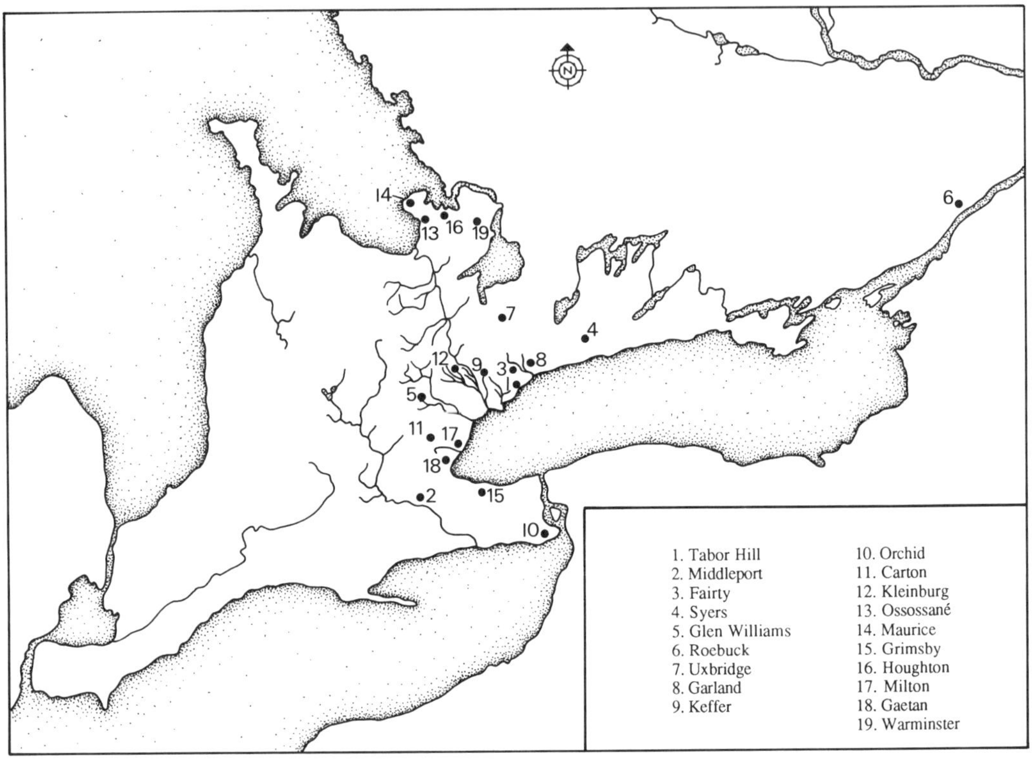

Fig. 1. Map of southern Ontario indicating the locations of the ossuary sites listed in Table 2.

into dietary shifts around the time of European contact. Based on carbon isotope values, maize was introduced between A.D. 600 and A.D. 1200; consumption reached a maximum around A.D. 1300, and it decreased slightly thereafter (Schwarcz et al., 1985). Katzenberg (1992) has compared carbon stable isotope results with data on bone infection, adult stature, and settlement patterns from four populations. She finds that the frequencies of bone lesions from infections (nonspecific periostitis and probable tuberculosis) are highest when maize consumption was highest (Fairty), and that they decline thereafter to comparatively low levels at historic Kleinburg. This could correlate with initial increases in population density and subsequent adaptation.

INFECTIOUS DISEASES

The diseases of the documented epidemics—measles, influenza, and smallpox—cannot be easily studied from skeletal remains. Although Jackes (1983) suggested that osteomyelitis variolosa consistent with smallpox can be identified in at least one Ontario ossuary, no subsequent work has been done on this topic. One important feature in establishing a differential diagnosis of smallpox is the bilateral distribution of bony lesions. This is very problematical in most ossuary contexts where right and left limbs cannot be matched. This commingling has also been a problem for scholars studying possible treponemal infections.

The specific infectious disease that has received the most attention in southern Ontario is tuberculosis. Tuberculosis affects the skeleton in a small proportion of cases, leaving changes that are distinctive enough to allow its identification with some confidence. Indeed, the ossuary burial pattern (large sample, all bony elements) assists our investigation of tuberculosis because the differential diagnosis applied to characteristic lytic lesions of the lower vertebrae is most reasonably predicated upon the frequency of the lesions within the population. If such lesions are very rare, they may represent occasional neoplasms. If they are found only among adults in a horticultural population such as the Huron, they may be caused by soil-borne fungi such as blastomycosis. However, if they are relatively common and found among all ages, tuberculosis is the most probable diagnosis.

Based on different researchers' identification of characteristic resorptive spinal lesions, there appear to have been numerous cases at precontact Fairty (Anderson, 1963), Glen Williams (Hartney, 1981), and Uxbridge (Pfeiffer, 1984), as well as contact Garland (Webb, 1969). Several other sites show occasional characteristic lesions, and we know of no ossuary that has been examined and has shown none.

Arguing that the Uxbridge ossuary shows an extremely high frequency of lesions, Pfeiffer (1984, 1986) has suggested that individual ossuaries may represent phases when host resistance differed through space and time. It appears highly probable that tuberculosis affected these populations since at least A.D. 1400. There is no evidence that postcontact sites show higher frequencies, as might be expected if population resistance were lowered through physical and psychosocial disruption. Perhaps we should expect lower lesion frequencies, based on the reasoning that severe disruption will cause mortality from competing causes of death or that compromised populations will experience higher mortality from the early nonskeletal phases of tuberculosis.

Nonspecific infections, often classified as osteomyelitis and periostitis, have been noted in all ossuaries studied to date. Clabeaux (1977) used the distribution of nonspecific periostitis to illustrate the problems of frequency calculation from ossuary data. The most commonly affected element in her study of the Orchid ossuary was the tibia, with 35/287 (12%) affected.

Subsequent studies can be used to illustrate the impact of variable diagnostic criteria on population comparisons. Larocque (1991) has attempted to assess temporal change in the frequency of periostitis. He examined subadult and adult right tibiae from three ossuaries: Fairty (precontact), Kleinburg (early contact), and Ossossané (late historic). Two samples, Fairty and Kleinburg, were also examined independently by Katzenberg (1992). Both researchers report high frequencies of periostitis, with Kleinburg showing slightly lower frequencies than Fairty (Table 3). Larocque,

TABLE 3. Nonspecific Periosteal Bone Responses of the Tibia, Percent Affected

	Fairty		Kleinburg		Ossossané
	K[a]	L[b]	K	L	L
N	581	225	944	327	237
Juvenile	57.3	84.6	31.1	75.6	97.1
Adult	78.1	100.0	65.1	93.8	97.2
Total	69.9	94.8	59.7	89.0	97.0

[a]Katzenberg (1992); N = left plus right tibiae.
[b]Larocque (1991); N = right tibiae.

TABLE 4. Nonspecific Periosteal Responses of the Tibia, Percent in Three Degrees of Involvement

		Fairty		Kleinburg		Ossossané
		K[a]	L[b]	K	L	L
Juvenile[c]	1	88		100		
	2	5		0		
	3	7		0		
Adult[c]	1	86	93.3	85	89.1	81.8
	2	5	6.7	6.5	6.6	13.4
	3	9	0	8.5	4.3	4.8

[a]Katzenberg (1992).
[b]Larocque (1991).
[c]Slight (1), moderate (2), and severe (3).

however, reports almost twice as many cases as Katzenberg. Despite these frequency differences among the adults the degrees of involvement observed (slight, moderate, severe) are similar (Table 4).

Based primarily on the assessment of active lesions, Larocque (1991) argues that inflammatory responses are more often severe at Ossossoné, and the healing process is more advanced in the Fairty sample. The Kleinburg sample is consistently closer to Ossossané than to Fairty. Larocque (1991) concludes that we may see here evidence of usual chronic infections exacerbated by postcontact epidemics. This scenario differs from that proposed by Katzenberg (1992). She cites higher frequencies of periostitis at Fairty as illustrative of an early village population adjusting to a diet high in maize and with high population density. The lower values at Kleinburg indicate "better" adaptation to high-density living. The comparison of these two studies highlights the pressing need for paleopathologists to calibrate their methods and illustrates the difficulty of interpreting pathology in skeletons.

TRAUMA

The study of trauma can help us confirm the fundamentally accident-prone nature of our species (Damon, 1977:146) and document interpersonal conflict. Healed fractures occur in ossuary remains; however, once again this interpretation is problematical. Anderson (1963) reports 11 healed fractures of the limbs from the 295 Fairty adults, plus a number of compression fractures of vertebrae, none specifically attributable to interpersonal conflict. Pfeiffer (1984) reports six cranial wounds, 22 infracranial fractures, and 21 vertebral compression fractures from 457 individuals from Uxbridge. Some wounds are clearly associated with interpersonal violence. Clabeaux (1977) does not work from a minimum population size for the postcontact Orchid site, but she does report two cranial, 27 axial, and 24 limb bone fractures. Seventeen of the axial fractures are vertebral compression fractures. Both skull fractures appear to be conflict related, and two bones have stone projectile points embedded in them. Interpersonal violence, therefore, predates contact, and Iroquoian fracture frequencies range from low at precontact Fairty to high at later sites. The numerous compression fractures may be due in part to rather poor quality cortical bone.

NUTRITIONALLY RELATED CONDITIONS

The nutritionally related assessment of precontact and postcontact ossuaries from southern Ontario has proceeded in earnest since the original investigation of Ossossané in the late 1940s. Disparate methods and the nonspecificity of traits contribute to interpretive difficulties.

Bone Mass Measures

Relatively low values for cortical and cancellous adult bone are reported by Pfeiffer and King (1983) in their study of the late precontact Uxbridge and postcontact Kleinburg samples. Mean values were slightly lower for Uxbridge. A more recent study using single-photon absorptiometry detected an apparent decrease in Iroquoian adult bone density over time with below normal measures appearing by the end of the fifteenth century (Southern, 1990). Whether these patterns are best explained by diet, diet-disease interaction, diagenesis, or some combination of factors cannot be adduced from the current data.

Growth Arrest Lines

Juvenile patterns of growth arrest based on radiographs of immature tibial diaphyses have been assessed in four ossuary samples (Table 5). The two precontact samples, Fairty and Uxbridge, have substantially lower frequencies than the two postcontact samples (41–45% versus 70%). However, juvenile representation varies considerably among these samples, which may introduce sampling error. Even if frequency differences are real, they are difficult to interpret. Were later juveniles getting sick more often, or were they more capable of recovering from ill health?

TABLE 5. Frequency of Growth Arrest Lines, Based on Tibia Diaphyses

Site	N[a]	%	Source
Fairty	47/114	41	Larocque, 1991
Uxbridge	30/66	45	Pfeiffer et al., 1986
Kleinburg	71/101	70	Larocque, 1991
Ossossané	26/37	70	Larocque, 1991

[a]Number with lines/total observed.

Cribra Orbitalia

Data on porotic hyperostosis (more specifically the orbital manifestation known as cribra orbitalia) are drawn from Larocque's (1991) study of the Fairty, Kleinburg, and Ossossané ossuaries. Data collected by a single researcher should minimize systematic error. Although other reports discuss porotic hyperostosis from additional sites (e.g., Hartney, 1978), Larocque's synthesis provides a more consistent picture.

The precontact Fairty ossuary is reported to have an overall cribra orbitalia prevalence of 19.6% (adults and juveniles, at least one orbit is scored as being affected). The Kleinburg ossuary has a prevalence of 26%, and that of Ossossané is 15%. These values are not significantly different. Females seem to be less affected than males at Fairty (15% as opposed to 22%) and Ossossané (13% as opposed to 17%). However, males and females are equal in the Kleinburg ossuary sample (24% compared to 25%). Larocque (1991) notes that differences arise in the young age groups and not in the 30 year and older age group. Figures for juveniles are 38% for Fairty, 9% for Ossossané, and 34% for Kleinburg.

These data suggest that cribra orbitalia was common among precontact juveniles, yet the value is not significantly different from postcontact Kleinburg. As discussed earlier, juvenile remains are likely to be underrepresented in ossuaries. Certainly, the Kleinburg Ossuary is purported to be deficient in very young children, precisely the group of interest when assessing a population for cribra orbitalia. Therefore, biased samples may cloud these results. Nevertheless, the Fairty juvenile figure (38%) is higher than at the two postcontact ossuaries, especially Ossossané. Thus, it seems that precontact young individuals were under more severe hematological stress of some type.

TABLE 6. Percent of Permanent Teeth with Dental Caries

Site	Incisors	Canines	Premolars	Molars	All
Fairty[a]	6	7	13	58	28
Orchid[a]	—	—	—	—	30.0
Glen Williams[b]	—	—	—	—	22.4
Roebuck[b]	—	—	—	—	25.1
Uxbridge[c]	4.4	6.9	15.7	32.3	20.9
Carton[a]	2.3	11.2	20.4	41.5	27.3
Kleinburg[b]	8.7	15.1	25.9	52.9	40.6
Maurice[d]	9.1	6.6	27.2	40.0	27.2
Gaetan[e]	2.0	15.6	25.7	51.8	32.8

[a]Hartney (1978).
[b]Patterson (1984).
[c]Gagné (1990).
[d]Jerkic (1975).
[e]Melbye et al. (n.d.).

Dental Caries

Dental caries frequencies provide insights into the subsistence base of past population. A large proportion of the dental caries data comes from Patterson (1984). He notes a very low frequency of dental caries for Early and Middle Woodland peoples. However, with the adoption of maize, there are substantial increases in dental caries frequencies by late Early Ontario Iroquois times, a trend that continues in the contact period. Various researchers have recorded data in different ways. The most frequently reported method, the frequency of the affected tooth type, is utilized here (Table 6). Frequencies are either provided in the original references or they are calculated from primary data.

Precontact Fairty shares a similar dental caries pattern with that of a probable contact ossuary, Orchid (28% and 30%, respectively). In fact, clearly postcontact ossuaries such as Carton and Maurice show similar frequencies to that of Fairty (27.3% and 27.2%, respectively). Kleinburg, however, shows a much higher overall frequency (40.6%) in permanent dentitions. One possible cause for this disparity is variability among investigations in the diagnostic criteria chosen for scoring a carious lesion as "present." Other sources of error include the proportion of teeth lost premortem, the preservational quality of the material, and the conservation of the research material. These data are generally consistent with Patterson's claim of an increase in dental caries over time. It is not clear that this trend was exacerbated by contact.

Alveolar Abscessing

A percentage of macroscopically observable alveolar abscessing for adult tooth sockets is presented for each ossuary where information is either recorded or can be calculated from raw data (Table 7). As with caries, the Kleinburg ossuary has the highest abscess frequency (12.7%). However, given the similar figures for the other ossuaries he studied, Patterson (1984:316) concludes that "alveolar abscessing cannot be used to differentiate among southern Ontario pre-Iroquois and Iroquois groups." Likewise, based on our expanded comparison, we cannot conclude that there is a disparity between precontact and postcontact peoples.

Enamel Hypoplasia

Enamel hypoplasia refers to macroscopic defects of the enamel surface. They may range from small surface pits to a linear ar-

TABLE 7. Percent of Adult Tooth Sockets with Alveolar Abscesses

Site	Incisors	Canines	Premolars	Molars	All
Glen Williams[a]	6.0	8.7	9.5	9.6	8.4
Uxbridge[b]	3.2	4.9	6.5	3.5	4.5
Orchid[a]	—	—	—	—	5.0
Kleinburg[c]	10.6	12.3	12.7	14.3	12.7
Maurice[a]	2.6	4.9	8.0	7.6	5.9
Milton[a]	2.1	3.0	3.5	7.9	4.7
Gaetan[d]	3.1	4.2	5.5	7.5	5.3

[a]Hartney (1978).
[b]Gagné (1990).
[c]Patterson (1984).
[d]Melbye et al. (n.d.).

rangement of several pits or linear furrows or grooves. These dental defects furnish a permanent record of physiological disturbances from birth to the final calcification of the third molar crown. The recording and interpretation of these lesions, however, is not without sources of error (Patterson, 1984).

Patterson (1984), Gagné (1990), and Larocque (1991) have examined adult dentitions from ossuaries for examples of enamel hypoplasia. Patterson's (1984) data describe two ossuaries, Glen Williams and Kleinburg (Table 8). The overall frequency of affected teeth is reported as 3.9% and 10.6%, respectively. Based on only these two sites, the most that could be concluded is that the frequency for the postcontact Kleinburg ossuary is over two and a half times more than that of the precontact Glen Williams ossuary. The data provided by Gagné for Uxbridge, with an overall frequency of 2.5%, are consistent with a scenario in which nonspecific stress affecting crown formation increases at contact. Alternatively, more contact children survived with defects in comparison to their early counterparts, who tended to die young.

Larocque (1991) has reassessed the Glen Williams, Kleinburg, and Ossossané ossuaries. This reassessment reveals a somewhat different trend than that indicated by Patterson's and Gagné's data. The Glen Williams and Kleinburg ossuaries show virtually identical values for the incisors (52.7% and 51.3%, respectively). Further, the affected

TABLE 8. Percent of Teeth with Enamel Hypoplasia

	Glen Williams		Uxbridge	Kleinburg		Ossossané
	P[a]	L[b]	G[c]	P	L	L
Incisors	3.5	52.7	1.2	26.3	51.3	41.7
Canines	17.3	61.6	16.1	50.9	52.8	53.2
Premolars	2.0		1.4	7.8		
Molars	0.6		0.3	5.0		
All	3.9	57.7	2.5	10.6	52.1	46.3

[a]Patterson (1984). Both maxillary and mandibular permanent teeth.
[b]Larocque (1991). Permanent maxillary central incisors, permanent mandibular canines.
[c]Gagné (1990). Both maxillary and mandibular permanent teeth.

incisor figure for Ossossané is significantly lower (41.7%) than those of the other two ossuaries (Larocque, 1991). The situation for the permanent lower canines is reported for the three ossuaries as 61.6% for Glen Williams, 52.8% for Kleinburg, and 53.2% for Ossossané. We hesitate to draw any firm conclusions from these studies insofar as they clearly conflict, and they serve as a reminder that research results are dependent on the research design and diagnostic criteria employed.

CONCLUSION

The Ontario ossuary samples have generally shown a young age at death, relatively poor cortical bone quality, dental disease of the type associated with a low wear gradient and a cariogenic diet, trauma from interpersonal conflict, notable amounts of nonspecific infection, symptoms of developmental nutritional deficiencies, and putative tuberculosis. These features are distinctive when compared with earlier nonhorticultural populations from the region. It appears that through precontact times the signs of ill health became more common. Perhaps cortical bone quality declines, dental caries frequencies increase, signs of nutritional deficiency increase, and tuberculosis remains common. However, any of these so-called trends could be influenced by sampling error, differential survivorship, or both. Finally, even when true changes can be documented, they are hard to interpret. For example, the two samples with the most tuberculous lesions, Glen Williams and Uxbridge, are from late in the precontact era. Because bony lesions from tuberculosis arise late in the progress of the disease, perhaps these two ossuaries actually represent the healthiest populations; these were people with enough resistance to survive initial bouts of infection that may have devastated other groups. The same interpretive dilemma can be extended to several other conditions.

A great deal of information is available, and more is potentially available from the thousands of individuals from ossuaries currently under curation. Trends that have been tentatively identified in limited skeletal series indicate that changes occurred over the several hundred years preceding European-derived disruption. With regard to European contact, definitive patterns have yet to emerge.

Perhaps we are asking questions that are too fine scaled. Despite the apparent richness of the available material, the extant ossuary samples represent a very small sample of a population estimated to have hovered between 28,000 and 30,000 in the sixteenth and early seventeenth centuries (Warrick, 1990). European-derived epidemics are known to have had a devastating population impact from 1634 to 1640, a very brief time. Any earlier European initiated biological changes would need to be demonstrated solely from bones. It is possible that carefully designed research using calibrated comparative criteria and numerous samples could contribute a clearer picture. However, it is also possible that such a mammoth effort might fail, given the large number of simplifying assumptions that would be required and the intergroup variability that would inevitably accompany such a broad-based sample.

Existing methods are invaluable for understanding large-scale temporal and geographic differences, but they are of limited use in elucidating short-term, abrupt changes, at least in this region. We have attempted to find consistent temporal changes that correspond with the timing of European contact over a large geographic area. The original diversity in the relative health status of populations across this region makes such a search unrealistic. Thus far, the most successful elucidations of possible European contact have focused on small-scale comparisons. As we try to paint fine-grained pictures for broad areas, our blurry edges detract from the resulting images.

REFERENCES CITED

Anderson JE (1963) The people of Fairty: An osteological analysis of an Iroquois ossuary. National Museum of Canada Bulletin, Contributions to Anthropology *193*:28–129.

Anderson JE (1972) Dawson site physical anthropology. In JF Pendergast and BG Trigger (eds.): Cartier's Hochelaga and the Dawson Site. Montreal: McGill-Queen's University Press, pp. 311–323.

Anonymous (1894–1895) Annual Archaeological Report, Ontario. Toronto.

Anonymous (1889–1890) Annual Archaeological Report, Ontario. Toronto.

Anonymous (1907) Annual Archaeological Report, Ontario. Toronto.

Archaeological Services Incorporated (1990) Manuscript on file, Toronto.

Biggar HP (1929) The Works of Samuel de Champlain, vol. 3. Toronto: The Champlain Society.

Birx HJ (1991) Craniometry of the Orchid Ossuary, Fort Erie, Ontario. Buffalo: Persimmon Press Monograph in Anthropology.

Churcher CS, and Kenyon WA (1960) The Tabor Hill ossuaries: A study in Iroquois demography. Human Biology *32*:249–273.

Clabeaux M (1967) The Paleopathology of the Orchid Site, Fort Erie, Ontario. Ph.D. dissertation, Department of Anthropology, State University of New York, Buffalo.

Clabeaux M (1977) Health and disease in the population of an Iroquois ossuary. Yearbook of Physical Anthropology *20*:359–370.

Cook P (n.d.) The Uxbridge ossuary. Interim report on the progress of the excavation sent to the Ontario Heritage Foundation. Manuscript on file, Ontario Heritage Foundation, Toronto.

Cybulski JS (1966) Demography of the Orchid Site, Fort Erie, Ontario. M.A. thesis, Department of Anthropology, State University of New York, Buffalo.

Damon A (1977) Human Biology and Ecology. New York: W.W. Norton.

Fagan BM (1991) Kingdoms of Gold, Kingdoms of Jade: The Americas Before Columbus. London: Thames & Hudson.

Gagné G (1990) Variations régionales de la pathologie buccale des Iroquoiens du sylvicole supérieur. Ph.D. thèse, Département d'Anthropologie, Université de Montréal.

Gruber JW (1958) Preliminary notes on Huron dentitions. Bulletin of the Eastern Archeological Federation *17*:90.

Harris RI (1949) Osteological evidence of disease amongst the Huron Indians. University of Toronto Medical Journal *27*:71–75.

Harris RI (n.d.) Notes on the paleopathology of the skeletal remains of Huron Indians from the ossuary at Cahigue opened in 1946. Unpublished manuscript, Department of Anthropology, University of Toronto.

Hartney PC (1978) Paleopathology of Archeological Aboriginal Populations from Southern Ontario and Adjacent Regions. Ph.D. dissertation, Department of Anthropology, University of Toronto.

Hartney PC (1981) Tuberculous lesions in a prehistoric population sample from southern Ontario. In JE Buikstra (ed.): Prehistoric Tuberculosis in the Americas. Evanston: Northwestern University Archeological Program, pp. 141–160.

Heidenreich C (1971) Huronia: A History and Geography of the Huron Indians, 1600–1650. Toronto: McClelland and Stewart.

Jackes M (1977) The Huron Spine: A Study Based on the Kleinburg Ossuary Vertebrae. Ph.D. dissertation, Department of Anthropology, University of Toronto.

Jackes MK (1983) Osteological evidence for smallpox: A possible case from seventeenth century Ontario. American Journal of Physical Anthropology *60*:75–82.

Jackes M (1986) The mortality of Ontario archaeological populations. Canadian Journal of Anthropology *5*:33–48.

Jackes M (n.d.) The osteology of the Grimsby site: A Neutral Indian cemetery of the early 17th century. Manuscript on file, Royal Ontario Museum, Toronto.

Jerkic S (1975) An analysis of Huron skeletal biology and mortuary practices: The Maurice ossuary. Ph.D. dissertation, Department of Anthropology, University of Toronto.

Kapches M (1979) Markham Middleport manifestations. Paper presented, Canadian Archeological Association, Vancouver.

Katzenberg MA (1992) Changing diet and health in pre- and proto-historic Ontario. MASCA Research Papers in Science and Archeology, University of Pennsylvania, vol. 9, pp. 23–31.

Katzenberg MA, and White R (1979) A paleodemographic analysis of the *os coxae* from Ossossané ossuary. Canadian Review of Physical Anthropology *1*:10–28.

Kenyon WA (1977) Some bones of contention: The Neutral Indian burial site at Grimsby. Rotunda *10*:4–13.

Kenyon WA (n.d.) The Grimsby site: A historic Neutral cemetery. Manuscript on file, Royal Ontario Museum, Toronto.

Kidd KE (1953) The excavation and historical identification of a Huron ossuary. American Antiquity *18*:359–379.

Kidd KE (1954) A note on the paleopathology of Ontario. American Journal of Physical Anthropology *12*:610–615.

Knowles FHS (1937) Physical Anthropology of the Roebuck Iroquois, with Comparative Data from Other Indian Tribes. National Museums of Canada Bulletin 87.

Larocque R (1991) Une étude ethnohistorique et paléoanthropologique des épidémies en Huronie. Ph.D. thèse, Département d'Anthropologie, Université de Montréal.

McIlwraith TF (1946) Archeological work in Huronia, 1946 excavation near Warminster. Canadian Historical Review *27*:394–401.

McIlwraith TF (1947) On the location of Cahiague. Transactions of the Royal Society of Canada *41*:99–102.

Melbye FJ (1967) An Analysis of the Infracranial Material from the Orchid Site. M.A. thesis, Department of Anthropology, State University of New York, Buffalo.

Melbye FJ (n.d.) The infracranial skeleton. In J.E. Anderson (ed.): The Osteology of the Orchid Site, Fort Erie, Ontario. Manuscript on file, University of Toronto.

Melbye FJ, Mullen GJ, and McFadden K (n.d.) The human osteology of the Gaetan site (AjGx-44). Report on file, Ontario Heritage Foundation.

Molto JE (1983) Biological Relationships of Southern Ontario Woodland Peoples: The Evidence of Discontinuous Cranial Morphology. National Museums of Canada Mercury Series, Paper 117.

Mullen GJ (1990) Human Osteology of the Warminster Site "BdGv-1" Ossuary. M.A. thesis, Department of Anthropology, University of Toronto.

Patterson DK (1984) A Diachronic Study of Dental Paleopathology and Attritional Status of Prehistoric Ontario Pre-Iroquois and Iroquois Populations. National Museums of Canada Mercury Series, Paper 122.

Pendergast JF (1975) The St. Lawrence Iroquois. Ontario Archaeology *25*:47–56.

Pfeiffer S (1980a) Spatial distribution of human skeletal material within an Iroquoian ossuary. Canadian Journal of Archaeology 4:169–171.

Pfeiffer S (1980b) Assignment of sex to adult femora from an ossuary population. Canadian Review of Physical Anthropology *1*:55–62.

Pfeiffer S (1983) Demographic parameters of the Uxbridge ossuary population. Ontario Archaeology *40*:9–14.

Pfeiffer S (1984) Paleopathology in an Iroquoian ossuary, with special reference to tuberculosis. American Journal of Physical Anthropology *65*:181–189.

Pfeiffer S (1985) Comparison of adult age estimation techniques using an ossuary sample. Canadian Journal of Anthropology *4*:13–17.

Pfeiffer S (1986) Morbidity and mortality in the Uxbridge ossuary. Canadian Journal of Anthropology *5*:23–31.

Pfeiffer S (1991) Rib lesions and New World tuberculosis. International Journal of Osteoarchaeology *1*:191–198.

Pfeiffer S, Katzenberg MA, and Kelley MA (1985) Congenital abnormalities in a prehistoric Iroquoian village: The Uxbridge ossuary. Canadian Review of Physical Anthropology 4:83–92.

Pfeiffer S, and King P (1983) Cortical bone formation and diet among protohistoric Iroquoians. American Journal of Physical Anthropology *60*:23–28.

Pfeiffer S, Stewart K, and Alex C (1986) Growth arrest lines among Uxbridge ossuary juveniles. Ontario Archaeology *46*:27–31.

Ramsden PG (1977) A Refinement of Some Aspects of Huron Ceramic Analysis. National Museums of Canada Mercury Series, Paper 63.

Ridley F (1947) A search for Ossossané and its environs. Ontario History *39*:7–14.

Saunders SR, and Melbye, FJ (1990) Subadult mortality and skeletal indicators of health in Late Woodland Ontario Iroquois. Canadian Journal of Archaeology *14*:61–74.

Saunders SR, Ramsden PG, and Herring DA (1992) Transformation and disease: Precontact Ontario Iroquoians. In JW Verano and DH Ubelaker (eds.): Disease and Demography in the Americas. Washington, DC: Smithsonian Institution Press, pp. 117–125.

Saunders SR, and Spence MW (1986) Dental and skeletal age determinations of Ontario Iroquois infant burials. Ontario Archaeology *46*:45–54.

Schwarcz HP, Melbye FJ, Katzenberg MA, and Knyf M (1985) Stable isotopes in human skeletons of Southern Ontario: Reconstructing paleodiet. Journal of Archaeological Sciences *12*:187–206.

Sirianni J (1967) The Orchid Site Dentition. M.A. thesis, Department of Anthropology, State University of New York, Buffalo.

Southern B (1990) An assessment of bone quality and age-related patterns of bone loss among

Iroquoian populations. Paper presented, Canadian Archaeological Association, Whitehorse, Yukon.

Stothers DM (1977) The Princess Point Complex. National Museums of Canada Mercury Series, Paper 58.

Stothers DM, and Metress JF (1975) A system for the description and analysis of pathological changes in prehistoric skeletons. Ossa 2:3–9.

Sundick R (n.d.) Pathology, a report prepared concerning the pathological features in a series of tibiae from the Carton Site. Unpublished manuscript, Department of Anthropology, University of Toronto.

Sutton RE (1988) Paleodemography and Late Iroquoian ossuary samples. Ontario Archaeology 48:42–50.

Thwaites RG (ed.) (1896–1901) Jesuit Relations and Allied Documents. 73 volumes. Cleveland: The Burrows Brothers Company.

Tooker E (1964) An ethnography of the Huron Indians 1615–1649. Bureau of American Ethnology Bulletin 190.

Trigger BG (1969) The Huron: Farmers of the North. New York: Holt, Rinehart & Winston.

Trigger BG (1976) The Children of Aataentsic: A History of the Huron People to 1660. Montreal: McGill-Queen's University Press.

Trigger BG (1985) Natives and Newcomers: Canada's "Heroic Age" Reconsidered. Montreal: McGill-Queen's University Press.

Warrick GA (1990) A Population History of the Huron-Petun, A.D. 900–1650. Ph.D. dissertation, Department of Anthropology, McGill University, Montreal.

Webb J (1969) The Garland Ossuary. M.A. thesis, Department of Anthropology, University of Toronto.

Webb J (1972) A Multivariate Analysis of Some Prehistoric and Historic Iroquois Crania. Ph.D. dissertation, Department of Anthropology, University of Toronto.

White M (1966) The Orchid Site Ossuary, Fort Erie, Ontario. New York State Museum Bulletin 38, pp. 1–24.

Winnicki P (n.d.) The Carton Site dentition. Unpublished manuscript, Department of Anthropology, University of Toronto.

Williamson RF (1990) Archaeological Services Inc.: Archaeological assessments in the Province of Ontario. Annual Archaeological Report, Ontario, vol. 1, Toronto: Ontario Heritage Foundation, pp. 72–76.

Wintemberg WJ (1948) The Middleport Prehistoric Village Site. National Museum of Canada Bulletin 109.

Wright JV (1966) The Ontario Iroquois Tradition. National Museum of Canada Bulletin 210.

Wright PJ (1974) The Dental Morphology of Three Ontario Iroquois Ossuary Populations. M.A. thesis, Department of Anthropology, McMaster University, Hamilton, Ontario.

Wrong GW (ed.) (1939) The Long Journey to the Country of the Hurons by Gabriel Sagard. Toronto: Champlain Society.

Chapter 6

Trade, Contact, and Female Health in Northeast Nebraska

Karl J. Reinhard, Larry Tieszen, Karin L. Sandness, Lynae M. Beiningen, Elizabeth Miller, A. Mohammad Ghazi, Christiana E. Miewald, and Sandra V. Barnum

Departments of Anthropology (K.J.R., S.V.B., C.E.M, K.L.S.) and Geology (A.M.G.), University of Nebraska, Lincoln, NE 68588; Department of Biology, Augustana College, Sioux Falls, SD 57197 (L.M.B., L.T.); and Department of Anthropology, Arizona State University, Tempe, AZ 85287 (E.M.)

Most scholars are understandably preoccupied with the impact of Europeans on native peoples who were passive, unwilling, or resistant participants in that contact. We present in this chapter a different case. The Missouri River tribes, including the Omaha and Ponca, willingly engaged in relations with Euramericans, especially in the fur trade that dominated interaction in this region.

The time frame for this study is 1780–1820, a period when interaction between individual traders and Native Americans was replaced by the dominance of the American Fur Company in organized exploitation of the Missouri River lands and peoples. This involvement later contributed to the well-documented environmental destruction of Plains river systems with dire consequence to Native American subsistence in the late 1800s. The years between 1780 and 1820 saw commerce contribute to the emergence of the Omaha tribe as a major political, economic, military, and cultural force in the area of modern Nebraska. It is thought that lifestyles changed during these dealings with Euramericans to the simultaneous benefit and peril of tribal members.

Omaha population decreased during the late 1700s and early 1800s. Epidemics (Trimble, 1989) and trade in toxic lead (Reinhard and Ghazi, 1992) contributed to this decline. Beyond these factors, female health and reproductive success must have played a role in the decline and eventual recovery of the Omaha population. It appears that several factors related to the fur trade resulted in changes in ways of life that adversely affected female health. Trade exposed the Omaha to Euramericans and their diseases. The need to produce furs resulted in greater physical demands on an already heavily taxed population. We suspect that the vitality of Omaha women was compromised, resulting in a decrease in life expectancy and reproductive success.

If extrapolations from ethnographic and historical writings are correct, these alterations in life experience should have left recognizable signatures on the skeletons. The comparison of precontact skeletons with contact period skeletons should illus-

In the Wake of Contact: Biological Responses to Conquest, pages 63–74

trate the general impact of contact. The comparison of contact male and contact female skeletons should indicate how the change in life experience affected the sexes. Our view of the contact period comes from the excavation of two Omaha cemeteries and one Ponca cemetery. The Omaha cemeteries (sites 25DK10 and 25DK2) are associated with the village of Tonwatonga (site 25DK5). These cemeteries yielded 92 individuals, including 37 adults (16 females, 15 males, two possible females, and four possible males). The village site was mentioned by Lewis and Clark, so it was occupied at the beginning of the nineteenth century. It may have been established as early as 1780, and it was probably abandoned by 1833. It is not mentioned by Maximilian in his Missouri River narrative (Thomas and Ronnefeldt, 1982).

We estimate that of 10,000 grave objects, listed under 1,300 catalogue numbers, 95% were of European or Euramerican manufacture. The artifacts suggest that the Omaha cemeteries were not used synchronously. Firearms and gun parts were more common in 25DK2, and they included powder horns, gun flints, gun parts, and a complete flintlock pistol. In contrast, two burials from 25DK10 contained only four musket balls. If firearm abundance increased over time, then the presence of more firearms at 25DK2 indicates a later use of the cemetery.

We have also initiated a study of a cemetery from Ponca Fort (25KX1). The Ponca tribe is related to the Omaha tribe, from which it split in the eighteenth century. Sixty skeletons were encountered, including 31 adults (26 women and seven men). In comparison with the Omaha, few trade goods were found. Ponca Fort was used contemporaneously with the Omaha sites.

ETHNOGRAPHIC BACKGROUND

Dorsey (1884) and Fletcher and LaFlesche (1911) described tasks performed by Omaha males and females. Primary among the female activities were duties associated with horticulture, maintenance of households, and hide preparation. Garden plots of one-half to three acres of land were devoted to maize, beans, squash, and tobacco. Simple horticultural tools were used, including bone hoes, rakes, and digging sticks. Metal tools were introduced later through trade. Food provided by efforts of the women contributed substantially to the tribe's diet and provided a surplus for trade to outside groups.

In addition to horticulture, most tasks related to sustaining households belonged to women, except for the provision of meat and the protection of the family. The construction and maintenance of the home were a woman's responsibility. They made and set up the tents that provided shelter during the hunting season and also often were used in villages. When enough resources were acquired for an earth lodge, both males and females participated in the construction of this more permanent dwelling (Fletcher and LaFlesche, 1911).

Both within and outside her home, an Omaha woman did the rest of the household chores. She tended to children, prepared food and clothing, gathered firewood, collected wild foods, and was practiced in quill and bead artwork. During the bison hunt, these tasks were accentuated by the frequent movement of the camp and the labor involved in its dismantling and establishment. Products from the hunt also required the labor and immediate attention of women. Meat was cut and jerky prepared before it spoiled, and hides needed to be dressed and tanned. The preparation of the hides of bison and other animals was a major female responsibility. This was true prior to Euramerican trade, and it increased during the contact period.

The barter in animal skins with the Euramerican traders promoted hunting for wealth in addition to subsistence. Work escalated for both sexes. When males increased their hunting time, females processed more pelts, a time-consuming and labor-intensive endeavor. Specifically, Fletcher and LaFlesche state that "in the old time one woman could scarcely give proper attention to all the skins secured by a good hunter; still less could she do the additional work occasioned by the pressure of trade" (1911:615).

Because of the workload demanded of both sexes, children were assigned tasks at early ages. This was especially true for young Omaha girls, who while young were trained in the preparation of skins, the making of tent covers, the sewing of garments, and all other household duties.

Thus, duties were assigned early in a woman's life and incorporated virtually all aspects of village society and family maintenance. The central position of women in the construction and maintenance of the household provided the foundation for the entire tribe's welfare. A weakening of this female infrastructure, either by social or biological disruption, would have had a pronounced effect on the survival of the Omaha. At least biological disruption is indicated in the ethnohistoric literature. Several epidemics in the late 1700s and early 1800s probably negatively impacted the Omaha village 25DK5 (see Trimble [1989] for a history of epidemics). These included three smallpox episodes (1778–1783, 1800–1803, 1836–1840) and whooping cough and cholera (1832–1834).

The ethnobotany of the Omaha has been studied in detail (Gilmore, 1913, 1977; Yanovsky, 1986; Kindscher, 1987). Table 1 illustrates a diversity of gathered and cultivated plant foods in the Omaha diet. Among the cultivated plants, the Omaha reportedly had 15 varieties of beans, eight varieties of squash, and many kinds of maize. They also cultivated watermelon, which was introduced directly to the Omaha with the fur trade in 1795 (Nasatir, 1990:492). There was also a diversity of wild plant foods reported to have been the dietary mainstay when cultivation was not possible. Tubers from *Glycine apios*, *Psoralea esculenta*, and *Sagittaria arifolia* were most important.

SKELETAL ANALYSES

To provide background for the increase of osseous pathology among the Omaha, especially women, a pathological study of historic and prehistoric skeletons was completed. The historic sample consisted of the Omaha and Ponca skeletons described above. The prehistoric skeletal sample consisted of 40 individuals from two prehistoric cemeteries near the Omaha cemeteries, including sites 25DK9 and 25DK13.

With regard to the vertebral column, several different types of pathological involvement were scored. Osteophytosis, also called vertebral lipping, refers to the degeneration of the centrum in response to age and activity. Osteoarthritis refers to the degeneration of the posterior synovial joints of the spine. Schmorl's nodes are depressions in the centrum resulting from the herniation of the intervertebral disks.

Vertebral Pathology

The goals of the vertebral analysis were to assess whether or not contact populations exhibited a higher prevalence of spinal pathology, and if so, whether this was attributable to an increased workload for women.

The study of vertebral pathology was limited to skeletons that had reconstructible vertebral columns. A comparison of the spines of 21 Omaha and Ponca with 18 prehistoric individuals was made with regard to prevalence and severity of two degenerative conditions related to activity, osteophytosis and osteoarthritis. A second comparison of vertebral pathology showed that trauma and mechanical stress were more characteristic of the contact period (Sandness and Reinhard, 1992).

A greater prevalence of osteoarthritis is evident in the precontact sample (17 of 18

TABLE 1. Common and Scientific Names of Plants and Fungus Eaten by the Omaha[a]

Common name	Scientific name	Part eaten
Corn smut	*Ustilago maydis* (DC.) Cda.	Spores
Morel mushrooms	*Morchella esculenta* (L.) Pers.	Fruiting body
Arrowleaf	*Sagittaria latifolia* Willd.	Tuber
Wild rice	*Zizania aquatica* L.	Caryopses
Maize	*Zea mays* L.	Kernels
Wild onion	*Allium canadense* L.	Bulb
Black walnut	*Juglans nigra* L.	Nut
Hickory nut	*Carya ovata* (Mill.) K. Koch	Nut
Hazelnut	*Corylus americana* Walt.	Nut
Hackberry	*Celtis occidentalis* L.	Nut
Yellow lotus	*Nelumbo lutea* (Willd.) Pers.	Tuber, seed
Wild gooseberry	*Ribes missouriensis* Nutt.	Fruit
Wild strawberry	*Fragaria vesca* L. var. *americana* Porter and *F. virginiana* Duchn.	Fruit
Wild raspberry	*Rubus occidentalis* L.	Fruit
Wild rose	*Rosa arkansana* Porter	Fruit (famine food)
Crab apple	*Pyrus ioensis* (Wood) Carruth	Fruit
Red haw	*Crataegus rotundifolia* Moench	Fruit
Juneberry	*Amelachier alnifolia* Nutt.	Fruit
Wild plum	*Prunus americana* Marsh.	Fruit
Chokecherry	*Prunus virginiana* L.	Fruit
Prairie turnip	*Psoralea esculenta* Pursh	Root
Indian potato	*Apios americana* Medic.	Tuber
Ground bean	*Amphicarpaea bracteata* (L.) Fern.	Bean
Garden bean	*Phaseolus vulgaris* L.	Bean
Soft maple	*Acer saccharinum* L.	Sugar
Elderberry	*Sambucus canadensis* L.	Fruit
Box elder	*Acer negundo* L.	Sugar
Milkweed	*Asclepias syriaca* L.	Greens
Ground cherry	*Physalis heterophylla* Nees	Fruit
Pumpkin, squash	*Cucurbita pepo* L.	Pulp, seed
Jerusalem artichoke	*Helianthus tuberosus* L.	Tuber
Watermelon	*Citrullus lanatus* (Thunb.) Mansf.	Melon
Buffalo berry	*Shepherdia argentea* Nutt.	Fruit

[a]Compiled from Gilmore (1913, 1977), Kindscher (1987), and Yanovsky (1986).

individuals). The severity of osteoarthritis is as follows: eight minor involvement (44.4%), five moderate involvement (27.8%), and four severe involvement (22.2%). Thirteen of 18 precontact individuals show evidence of osteophytosis. With regard to severity, seven of the 18 (38.8%) show minor osteophyte development, two (11.1%) show moderate development, and four (22.2%) show severe osteophytosis.

Fifteen of the 21 Omaha/Ponca individuals display osteoarthritis with the following severity: eight (38.1%) minor involvement, six (28.6%) moderate involvement, and one (4.8%) severe involvement. Osteophytosis is evident in 11 of 21 historic people. Severity is as follows: minor in two (9.5%), moderate in five (23.8%), and severe in four (19%) individuals.

The greater prehistoric prevalence in osteoarthritis is largely due to longer average life (by about 15 years) in prehistory in comparison to the Omaha and Ponca (Sandness and Reinhard, 1992). The severity of osteoarthritis is greater and is also related to higher age at death for precontact individ-

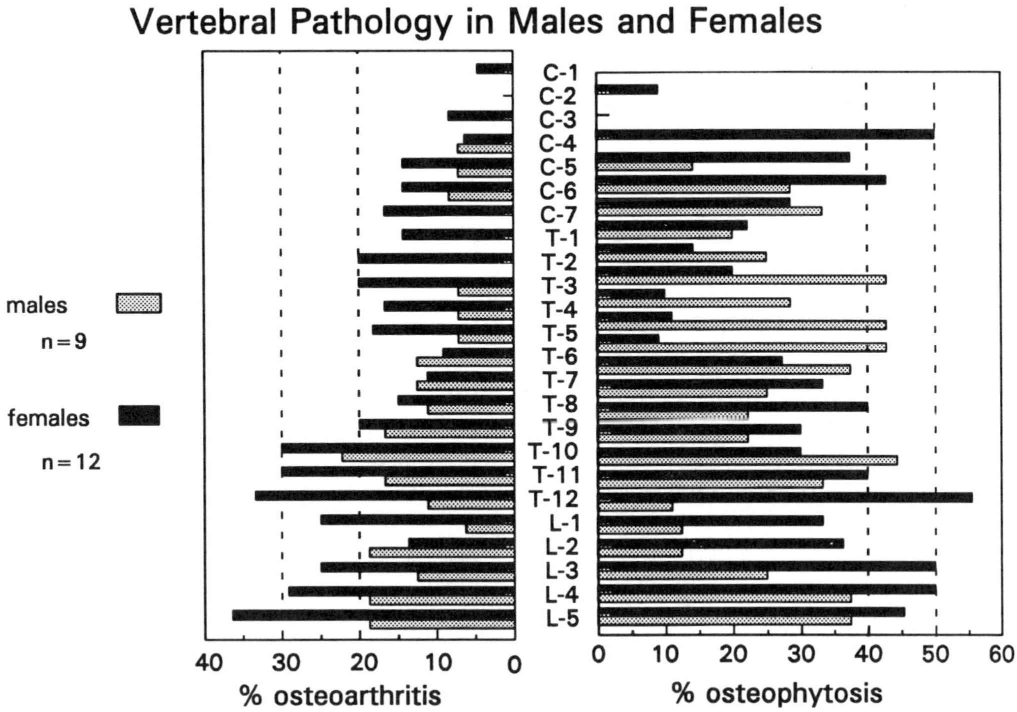

Fig. 1. Distribution of degenerative disease in male and female vertebral columns.

uals. Prevalence of osteophytosis is greater among the precontact sample, but severity is greater among the Omaha and Ponca. Osteophytosis is a more sensitive indicator of activity-induced stress (Sandness and Reinhard, 1992), and therefore the increased severity indicates that the spines of the historic people were under greater activity-related stress.

Evidence of trauma or mechanical stress in the Nebraska vertebrae increases after contact. Schmorl's nodes are found in three of 18 prehistoric people (16.7%) in contrast to nine historic people (42.8%). Spondylolysis is present in five historic adults (23.8%), but there is no evidence of the condition in prehistoric people.

Vertebral Pathology of Males and Females

Spondylolysis and Schmorl's nodes were used as indicators of mechanical stress and trauma-induced activities. Twelve female spines and nine male spines from the Omaha and Ponca were complete enough for study (Sandness and Reinhard, 1992).

Apparently, females were engaged in different activities than males (Fig. 1). Osteoarthritic vertebral degeneration is more frequent in females (75%) than in males (66.6%), and females exhibit higher frequencies of moderate (33.3%) and severe (8.3%) stages. In contrast, males are more commonly affected with osteophytosis (66.6%) than females (50%). The women exhibiting osteophytosis are moderately or severely affected in comparison to men who show primarily minor involvement. Both conditions affect males primarily in the lower thoracic and lumbar regions, especially at the tenth thoracic vertebra. Females exhibit higher frequencies in the upper thoracic and lower lumbar vertebrae. Overall, women also have a greater pathological involvement throughout the spine than men.

The distribution of osteophytic vertebrae is different between men and women. Female spines exhibit frequencies of 40% or more in the upper and mid-cervical (C-3 and C-5), middle and lower thoracic (T-8, T-11, and T-12), and lower lumbar vertebrae (L-3 to L-5). In contrast, male spines display frequencies of 40% or more in the upper and middle thoracic vertebrae (T-2, T-4, T-5, and T-10).

Schmorl's nodes are present in three women, but six of nine males have the condition. Of these, five males have more than one vertebra with a Schmorl's node, whereas only one woman has multiple Schmorl's nodes. Thus, it is possible that herniation of the intervertebral disks was more of a problem for men than women.

Among the Omaha, trauma is evidenced by four examples of spondylolysis (three affecting L-5). In every instance, there is complete bilateral separation in the region of the pars interarticularis. Three cases are women, all of whom were less than 25 years old at death (25% of females). The only male with the condition died at about 40 years of age (11% of males). It is noteworthy that one female with spondylolysis also has bilateral stress fractures of the first rib. In clinical settings, such fractures usually occur in people who habitually lift and carry heavy loads (Phoebe Kaplan, University of Nebraska Medical Center, personal communication).

In general, it seems that women were more profoundly affected by degenerative disease of the spine than men. This observation is consistent with ethnographic data (Fletcher and LaFlesche, 1911). Spondylolysis in women may have resulted from the stooped-over posture during hide scraping as depicted by Fletcher and LaFlesche (1911:Fig. 71).

Role of Horse Riding, Trauma, and Degenerative Disease

Skeletal samples from the Omaha (25DK2 and 25DK10) and Ponca (25KX1, 25KX5, and 25KX9) sites were examined for indications of horseback riding (Miller, 1992). Habitual riding is indicated by the presence of superior elongation of the acetabulum, extension of the articular surface of the femoral head onto the femoral neck, and enlarged muscle attachments for the gluteus medius and gluteus minimis, the adductor magnus and brevis, the vastus lateralis, and the medial head of the gastrocnemius muscles. Osteoarthritis of the first metatarsal, possibly from the use of toe stirrups (a thong through which the first toe was thrust), was also considered.

For sex comparison, the samples were divided into two activity groups, including those with (riders) and those without (nonriders) the osteological characteristics of horseback riding. The rider sex distribution is five females and 13 males, and the nonrider distribution is eight females and six males. Distribution by age groups, including adolescents (14–19), young adults (20–29), middle adults (30–39), and old adults (40+), show that individuals classified as both riders and nonriders are represented in all age groups for both sexes (Table 2).

Of 14 nonriders, there are four cases of vertebral mechanical stress, all of which are examples of spondylolysis in females. This spondylolysis in females may not be associated with horseback riding, but with hide scraping, as noted above. Schmorl's nodes are associated with one nonrider. In addition, there are three cases of trauma, including one fractured ulna, one dislocated clavicle, and one example of blunt trauma to the lower leg. Only one of these individuals exhibits vertebral degenerative changes. This individual is an old female, and degenerative conditions were expected regardless of riding status.

The group of riders (18 individuals) shows a much different picture of skeletal trauma and degeneration. There are 10 cases of degenerative disease excluding the metatarsal, 10 cases of trauma, and 12 cases of first

TABLE 2. Vertebral Mechanical Stress, Skeletal Trauma, and Osteoarthrosis of First Metatarsal Among Riders (5 Women, 13 Men) and Nonriders (8 Women, 6 Men)

	Adolescent		Young adult		Middle adult		Old adult	
	Male	Female	Male	Female	Male	Female	Male	Female
Nonriders								
Mechanical stress	0	2	0	1	0	0	0	1
Trauma	0	0	1	0	1	0	0	1
Osteoarthrosis	0	0	0	0	0	0	0	1
Riders								
Mechanical stress	0	1	3	1	2	0	2	1
Trauma	0	0	3	1	2	1	2	1
Osteoarthrosis	1	0	3	1	3	1	2	1

metatarsal arthritis in this sample (Table 2). Spondylolysis is present among two male riders (one of these had an incomplete spine and therefore was not noted in the vertebral analysis). Other forms of degenerative disease and mechanical stress include fusion of vertebrae and long bones, "kissing" lumbar spines, which indicate hyperflexion of the lumbar spine, Schmorl's nodes, and atrophy of a shoulder following a fracture of the humerus. Trauma is represented in seven individuals as fractures, in one individual as blunt trauma, and in one person as a knife wound to the forehead (this skeleton also has a dislocated shoulder). The occurrence of joint degeneration is also high among horseback riders of all age groups and both sexes. Twelve individuals exhibited degenerative changes in the joints, primarily the vertebrae, knees, hips, feet, and elbows.

Female trauma was not related to horseback riding. There are no significant differences between female riders and nonriders (χ^2, $p > 0.05$). Female riders show a younger age of onset for all conditions. Significant differences are present, however, between male riders and nonriders in mechanical stress frequency (χ^2, $p > 0.25$) and osteoarthrosis (χ^2, $p > 0.05$), but not in the frequency of trauma (χ^2, $p > 0.05$). These findings indicate that the degenerative disease and trauma present in men was related to a primary male activity: horseback riding. Such conditions in females were related to a rigorous lifestyle, unrelated to riding, that would have included strenuous labor while planting crops, foraging, burden carrying, and preparing hides.

Diet

Dietary reconstruction was completed through stable nitrogen and carbon isotope analysis in order to determine whether dietary differences existed between Omaha men and women, and whether such differences, if found, could be implicated in women's health (see Schwarcz and Schoeninger, 1991). Plants gathered by the Omaha have a C_3 metabolic pathway. Maize was the dominant C_4 plant that was consumed.

The carbon stable isotope data for the Omaha and Ponca are plotted in Figure 2. The carbon to nitrogen ratios of all values falls between 3.0 and 3.1, which indicates that preservation of collagen is good. When the values are examined by cemetery, there seems to be a difference between male and female diets at 25DK2, but not at 25DK10. Perhaps women at 25DK2 consumed more maize than males. This suggests that dietary habits changed between the use of the cemeteries. Among the Ponca, there is no difference between male and females values

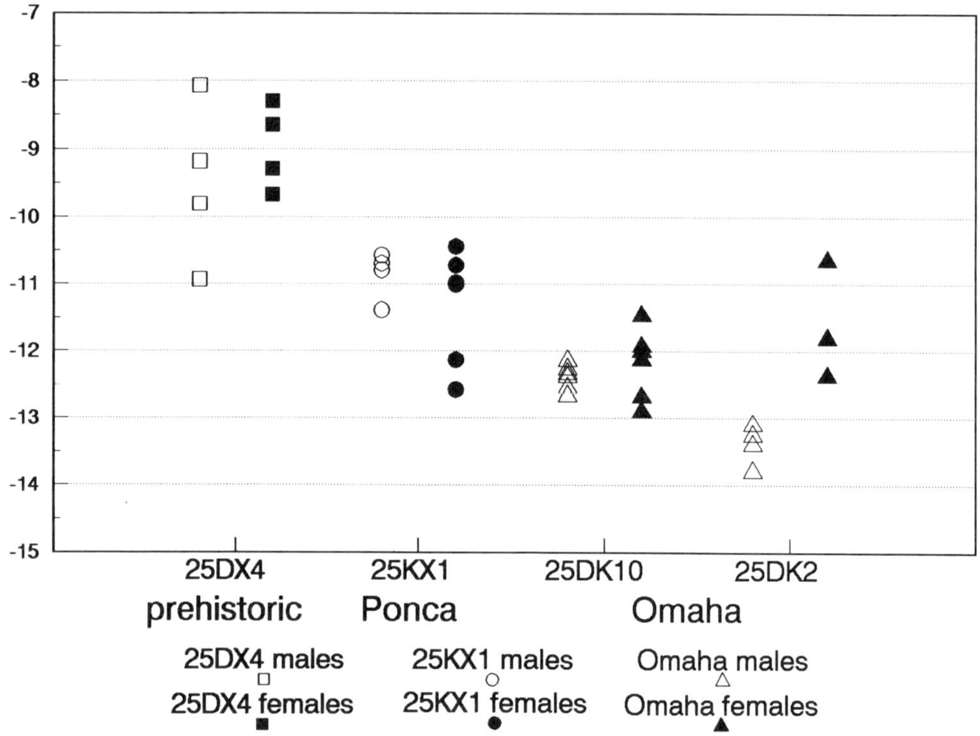

Fig. 2. Delta carbon values for Omaha and Ponca skeletons.

(Fig. 2). For all three sites, the males exhibit less variation than females, reflecting less dietary variability in males than females.

In Figure 3, the stable carbon and nitrogen isotope values for Omaha males and females are plotted against prehistoric St. Helena Phase horticulturalist skeletons from cemeteries near the Omaha cemeteries. The values for the St. Helena peoples fall within the range for maize horticulturalists. The Omaha tend to have more negative carbon values and more positive nitrogen values. Overall, it seems that Omaha diet included more animal protein or gathered plants. Presumably, then, the introduction of horses and firearms to the Omaha resulted in more successful hunting and greater access to meat as well as more effective foraging for wild plant foods.

Stress Associated with Childbearing

Among the Omaha remains, three fetal skeletons were identified. One was found in the abdominal cavity of a female aged 15.5–17.5 years, an individual who died during pregnancy. The other two fetal skeletons, presumably from miscarriages or stillbirths, were not associated with adult female remains. The presence of these remains suggests that spontaneous abortion and death during pregnancy were metabolic strains on Omaha women. It is possible that the stress of pregnancy contributed to the death of the pregnant woman noted above in combination with physical demands on young Omaha females as indicated by spinal pathology and ethnographic references.

Dental Wear

Barnum and Peterson (1991) have completed an analysis of dental wear for pre-

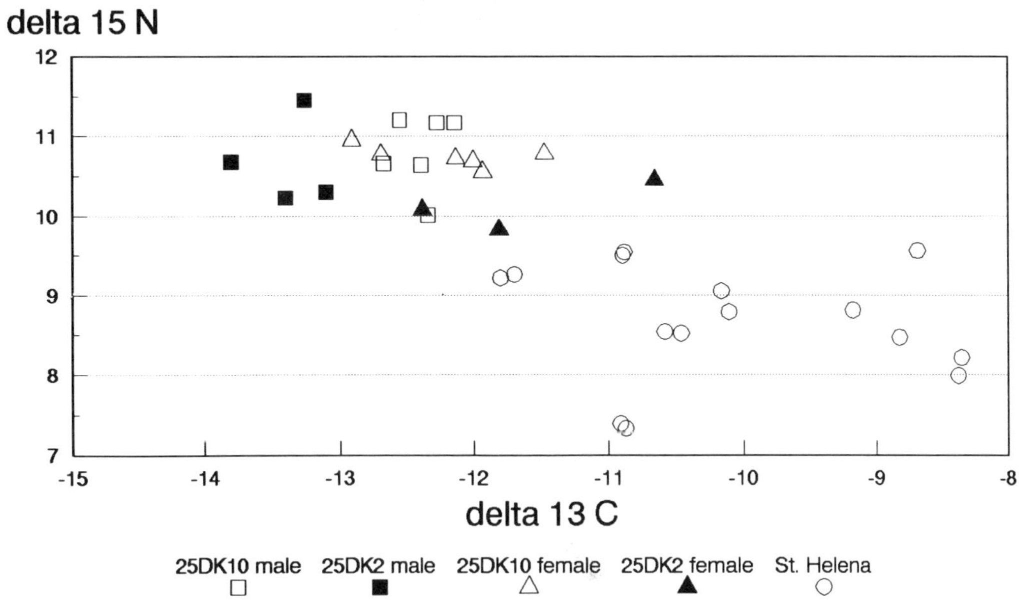

Fig. 3. Carbon and nitrogen stable isotope values for Omaha and prehistoric skeletons.

historic and historic native Nebraskans. Wear on 32 prehistoric and 14 Omaha dentitions was classified on a five-point system: 0, no wear; 1, wear of enamel only; 2, initial exposure of dentin; 3, extensive exposure of large areas of dentin; and 4, dentin completely exposed (Fig. 4). Omaha females show greater attrition than males, and the Omaha have more heavily worn teeth than the prehistoric inhabitants of 25DK13 and 25DK9 (Barnum and Peterson, 1991).

Forty-eight percent of all teeth exhibit level 4 wear. The greatest wear is on the anterior teeth (83% exhibit severe wear). Females have a greater percentage of severely worn teeth (61%) than males (43%), even though Omaha females generally died earlier in life than males. In contrast, only 15% of precontact teeth are severely worn. Again, severe wear is more common on the anterior teeth: 29% of the anterior teeth fall into this category, and 8% of the posterior teeth are similarly worn. With regard to sex, 21% and 7% of the precontact male and female teeth, respectively, are severely worn.

These wear patterns probably reflect the use of these teeth as tools for manipulating or holding objects. Omaha women, in particular, had pronounced anterior tooth attrition. Because tooth wear leads to pulp exposure and abscessing, increased dental wear among Omaha women likely predisposed them to dental disease. The cause of increased anterior wear in female dentitions might be related to softening hides by chewing.

Trade in Toxic Metals

A study of lead isotopes in these bones clearly demonstrates that lead from several geological sources were traded to the Omaha (Reinhard and Ghazi, 1992). Metallic lead was traded from present-day Missouri, and it was manufactured into ornaments and musket balls by the Omaha. Some Omaha skeletal lead signatures indicate absorption of lead from this source. There was, however, a more insidious source of lead—a red pigment used as a cosmetic. The isotope analysis of pigment indicates that this lead

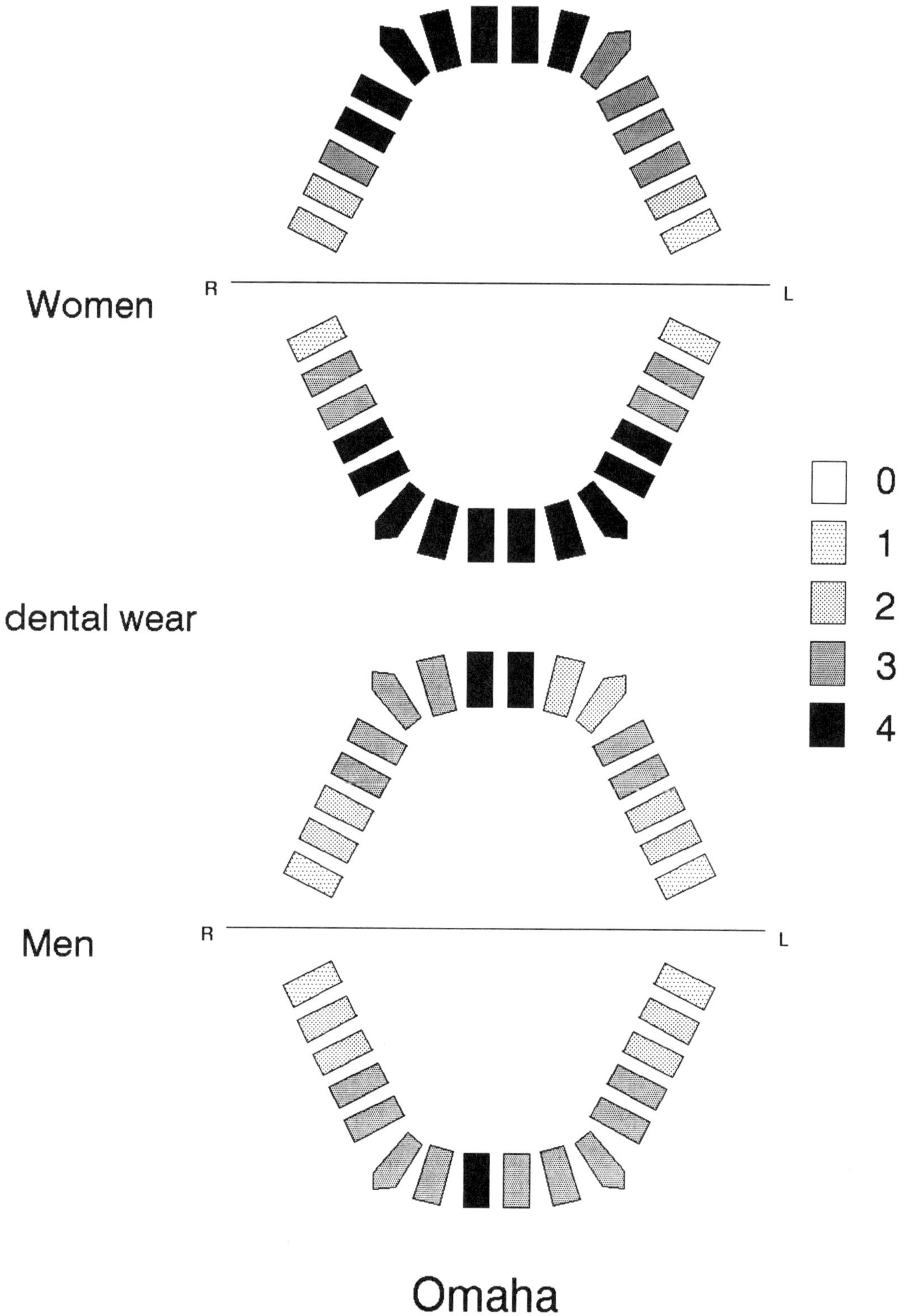

Figure 4.

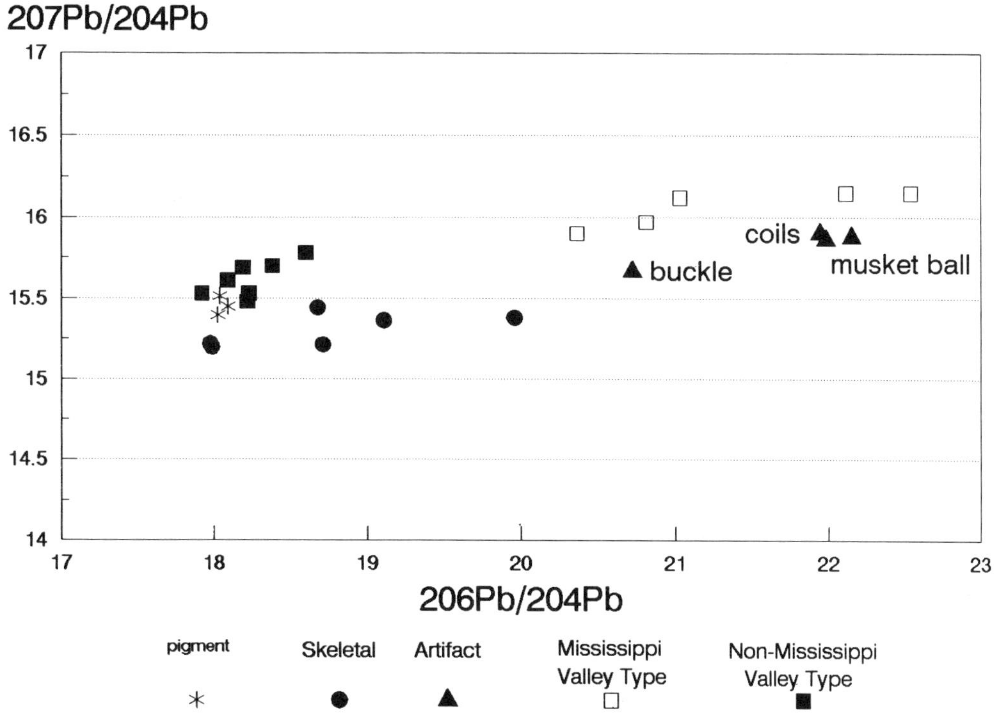

Fig. 5. Lead isotope values for skeletons, artifacts, and pigments in comparison to mine values from Mississippi Valley type deposits and non-Mississippi Valley type deposits.

came from a non-Mississippi type deposit, possibly from the northeastern United States. Most of the lead in the Omaha bones is derived from this source (Fig. 5). Evaluation of the lead levels indicates that most of the values are due to diagenetic absorption of lead from the pigment. Those values that cannot be attributed to this source are low (below 60 mg/g). The low levels may have been debilitating but were not lethal.

CONCLUSION

The analysis of Omaha and Ponca skeletons indicates that there were positive and negative effects of the fur trade. The main positive impact of the trade was the introduction of firearms that, when combined with horse riding, resulted in more efficient hunting practices and greater access to animal sources of protein. It also allowed efficient foraging for wild plants (Table 1) to supplant agricultural foods. Thus, in comparison to their prehistoric Nebraska counterparts, the Ponca and Omaha diet was probably more nutritionally sound.

Other analyses suggest a negative impact of the fur trade on the lives of the Omaha, in general, and particularly women. There is an earlier onset of degenerative disease among the Omaha in comparison to precontact peoples. The Omaha and Ponca male and female skeletons indicate that heavy labor associated with the fur trade combined with traditional female tasks led to more pronounced degenerative disease, especially among females. Stress during

Fig. 4. Distribution and severity of tooth wear for Omaha males and females. Mean score for tooth wear is presented. See text for explanation.

early years of pregnancy is suggested by the presence of fetal skeletons and the remains of a pregnant woman in the contact period cemeteries. The fact that the pregnant woman was 15.5–17.5 years old at death suggests that early pregnancy was an aspect of female life for the Omaha. Horseback riding seems to be a primary activity for men, and one of many activities carried out by women. Thus, the general depiction of male and female activities of men and women by Fletcher and LaFlesche (1911) is supported by the skeletal analysis presented in this chapter.

Increased anterior dental wear is associated with the Omaha in comparison to precontact peoples; it is especially characteristic of Omaha women. These observations suggest that the activities resulting in tooth wear increased in historic times, probably due to processing goods (e.g., furs) for trade. Such tasks placed women at greater risk to other dental diseases and, hence, lowered the quality of their lives.

ACKNOWLEDGMENTS

The studies summarized in this paper were jointly sponsored by the Omaha tribe of Nebraska and the University of Nebraska. The work was done at the request of the Omaha tribe, which recognized the medical/cultural value of skeletal study. Funding for the analysis was provided by the Vice Chancellor for Research, the University of Nebraska State Museum, the Department of Anthropology, and the University of Nebraska Dental School.

REFERENCES CITED

Barnum SV, and Peterson JK (1991) Division of labor by sex as seen in dental wear. Paper presented at the 49th meeting of the Plains Anthropological Association, Oklahoma City.

Dorsey JO (1884) Omaha Sociology. Bureau of American Ethnology 3rd Annual Report, pp. 205–370.

Fletcher A, and LaFlesche F (1911) The Omaha Tribe. Lincoln: University of Nebraska Press.

Gilmore M (1913) A study in the ethnobotany of the Omaha Indians. Journal of the Nebraska State Historical Society 17:314–357.

Gilmore M (1977) Uses of Plants by the Indians of the Missouri River Region. Lincoln: University of Nebraska Press.

Kindscher K (1987) Edible Plants of the Prairie. Lawrence: University of Kansas Press.

Miller E (1992) The effects of horseback riding on the human skeleton. Paper presented, Paleopathology Association, Las Vegas.

Nasatir AP (1990) Before Lewis and Clark. Lincoln: University of Nebraska Press.

Reinhard KJ, and Ghazi AM (1992) Evaluation of lead concentrations in Nebraska skeletons using ICP-MS. American Journal of Physical Anthropology 89:183–195.

Sandness KL, and Reinhard KJ (1992) Vertebral pathology in prehistoric and historic skeletons from northeast Nebraska. Plains Anthropologist 37: 299–309.

Schwarcz HP, and Schoeninger MJ (1991) Stable isotope analyses in human nutritional ecology. Yearbook of Physical Anthropology 34:283–321.

Thomas D, and Ronnefeldt K (1982) People of the First Man. New York: Promontory Press.

Trimble MK (1989) Infectious disease and the northern Plains horticulturalists: A human behavior model. Plains Anthropologist 34:41–59.

Yanovsky E (1986) Food plants of the North American Indians. In RI Ford (ed.): An Ethnobotanical Source Book: The Uses of Plants and Animals by American Indians. New York: Garland Publishing.

Chapter 7

Culture Change, Demographic History, and Health and Disease on the Northwest Coast

Jerome S. Cybulski

Archaeological Survey of Canada, Canadian Museum of Civilization, Hull, Quebec J8X 4H2, Canada

Native peoples of the Northwest coast first came into contact with Europeans late in the eighteenth century. Historical documents and related reconstructive techniques indicate that the next 100 years of intense contact with Europeans and other foreigners saw significant changes in native culture, demography, health, and disease (Duff, 1964; Cole and Darling, 1990; Boyd, 1990). In this chapter, these topics are viewed from the perspectives of human osteology and mortuary archaeology, and prehistory is included in the discussion, a significant 5,000+ years of Indian history prior to white influence. I consider positive and negative evidence for change through time, focusing on the coast of British Columbia.

The data are drawn from two recent studies wherein detailed references and other sources of information may be found (Cybulski, 1990, 1992). For different purposes, I provided overviews of what is currently known about human remains from the Northwest coast with respect to chronology, mortuary practices, vital statistics, paleopathology, and human biology. Some of the skeletal samples have been reorganized for the present purpose, but the methods for their analyses remain unchanged.

The Northwest coast culture area of native North America extends from the Copper River Delta in the Gulf of Alaska to the Oregon-California border, bounded inland by the Chugach and Saint Elias mountain ranges of Alaska, the Coast Mountains of British Columbia, and the Cascade Range of Washington and Oregon (Suttles, 1990). Human skeletal remains and archaeological data are best known from the Canadian segment which has traditionally represented classic Northwest coast culture in the writings of most anthropologists starting with Franz Boas in the nineteenth century (Adams, 1981).

THE SKELETAL SAMPLES

Historic Skeletal Remains

Ethnographers portray the native people of the Northwest coast in terms of discrete local or regional populations founded on differences in language, social organization, material culture, ceremonial attributes, polity, and economy. To a large extent, the cultural diverseness and territorial boundaries seen in the eighteenth and nineteenth centuries are maintained by the descendants of these earlier peoples. Major groups in British Columbia include Haida on the

In the Wake of Contact: Biological Responses to Conquest, pages 75–85

Queen Charlotte Islands; Tsimshian on the northern mainland coast, encompassing Nisga'a, Gitksan, Coast Tsimshian, and Southern Tsimshian; Haisla, Haihais, Heiltsuk (Bella Bella), and Oowekeeno on the central mainland coast (collectively northern Kwakiutl in the past); Nuxalk (Bella Coola), a nearby Salishan language isolate; Kwakwaka'wakw and Kwa-giulth (collectively southern Kwakiutl) on the north and east coasts of Vancouver Island and the adjacent mainland; Nuu-chah-nulth (northern and central Nootka) and Nitinaht on the west coast of Vancouver Island; and various Salishan (Coast Salish) subdivisions about the Strait of Georgia (see Suttles [1990] for territorial boundaries and nomenclature).

Major museum expeditions from New York and Chicago in the 1890s, as well as lesser excursions thereafter by other institutions and individuals, produced sizable collections of human skeletal remains from coastal British Columbia (Cybulski, 1990). Chronologically related skeletal remains have also been produced by controlled archaeological excavations in the 1960s and 1970s (Cybulski, 1992). Although information is spotty for some museum-collected bones, most provenience data indicate the remains are those of people who died about the time of initial European contact up to, and including, the last decades of the nineteenth century. In total, 593 individuals are currently known from 47 historic period mortuary localities (Cybulski, 1992). Almost all regions of the British Columbia coast are represented, a major exception being Tsimshian territory.

Prehistoric Skeletal Remains

British Columbia coast prehistory has been divided into four stages: Lithic (8000 B.C. to 3500 B.C.), Early Developmental (3500 B.C. to 1500 B.C.), Middle Developmental (1500 B.C. to A.D. 500), and Late Developmental (A.D. 500 to European contact, ca. A.D. 1774) (Fladmark, 1986). Although the stages may be applied to the entire coast, separate regional histories, sometimes with culture phase or type designations, have been developed for the Queen Charlotte Islands, Prince Rupert Harbour on the northern mainland coast, the central mainland coast about the site of Namu, and the south coast Strait of Georgia (Fladmark, 1986).

With the possible exception of one or two individuals, human remains and mortuary sites are unknown prior to 3500 B.C. This time marks the beginning of the shell midden buildup characteristic of the vast majority of known prehistoric sites on the British Columbia coast. The remains of a minimum 759 skeletal individuals have been excavated from 30 such sites (Cybulski, 1992). The figures only reflect site data with age-at-death and gender information based either on my analyses or the reports of other investigators. At least 12 other midden sites with human burials are known, including some 250 additional skeletons. Those data are incomplete and only partly acknowledged in this study, but it is not likely that their full inclusion would materially alter the known prehistory stage findings (see Cybulski, 1992).

Included among the 759 individuals are 288 from 10 closely situated sites at Prince Rupert Harbour (traditionally Coast Tsimshian territory) and 342 individuals from 17 sites in the Strait of Georgia region (traditional Coast Salish territory). They represent the two largest regional assemblages of known prehistoric remains from the British Columbia coast. Other cultural entities are represented by single mortuary sites at Greenville in the Nass River valley north of Prince Rupert Harbour (traditional Nisga'a territory), Blue Jackets Creek on the Queen Charlotte Islands (traditional Haida territory), and Namu on the central mainland coast (traditional Heiltsuk territory).

Chronology of Human Remains

The temporal distribution of known human remains is notably uneven. The Early and Late Developmental Stages (77 and 84 individuals, respectively) account for only 12% of the total sample that includes the historic period. Those of the Early Stage are largely from Blue Jackets Creek (28) and Namu (33), and those of the Late Stage are largely from Greenville (59). The Middle Developmental Stage accounts for most prehistoric remains (584), including 284 at Prince Rupert Harbour and 293 at Strait of Georgia. Except for Greenville, which is entirely Late Stage, some remains from the other entities have been assigned to other stages (seven Middle and two Late from Namu, four Late from Prince Rupert Harbour, 16 Early and 19 Late from Strait of Georgia).

CULTURAL CONTEXT

Most modern archaeological studies in British Columbia emphasize cultural continuity from the Early Stage to the time of European contact and promote an inverted pyramid approach to interpretations of cultural development and population growth (e.g., Mitchell, 1971; MacDonald and Inglis, 1981; Burley, 1989). For the most part, only gradual changes are recognized in terms of artifactual evidence, building on earlier themes. Interpretations of prehistoric social, economic, political, and ideological content draw heavily on Northwest coast ethnographies, producing circular arguments in support of the archaeological thesis of cultural continuity. In the overall scheme, only European contact caused significant changes, though historians indicate that economic changes of the time were flexibly adaptive rather than abrupt and fully disruptive (Cole and Darling, 1990). Mortuary and human skeletal evidence, however, suggest important cultural changes in the prehistoric period involving methods of corpse disposal, intentional head shaping, and the use of labrets or lip plugs.

Methods of Corpse Disposal

The paucity of known Late Developmental Stage remains may reflect a significant change in methods of corpse disposal before white contact. All known historic period human remains are from above-ground mortuary localities, consistent with ethnographic observations that point solely to above-ground disposal of the dead (Brabant in Moser, 1926:59; Collison, 1981:65; Cybulski, 1992). Specific methods varied locally and included placing bodies in boxes in trees, in rock crevices, rock shelters, or caves, in constructed grave houses, on surface or scaffold-like wooden platforms, or on elaborately carved "totem" poles raised to commemorate the death of a chief. When the epidemics brought by European-borne infections were at their worst, some Queen Charlotte Islanders simply placed their dead in boxes and left them outside their lodgings (Collison, 1981:65).

Although below-ground burial was encouraged by Christianity, it also appears to have been a preferred method of corpse disposal in the distant past. With one possible, but unverified, exception (Skinner, 1989; Cybulski, 1992:31), all known prehistoric mortuary localities are shell midden sites. Human remains are most often found as single articulated skeletons in shallow, but occasionally deep, pits dug specifically for the purpose. The evidence for burial pits is both definite and circumstantial, either structurally evident in the middens or suggested by later burials dug through earlier ones (Cybulski, 1992).

Disarticulated and scattered human bones are also found in shell middens, and some archaeologists have speculated that they are the disturbed remnants of above-ground corpse disposal methods, thereby demonstrating long-standing prehistoric cultural continuity with the historic period (e.g., Burley, 1989). However, the human elements are almost always found in the same areas of the middens as intact burials and are more likely the remnants of disturbed, formerly intact interments (Cybulski, 1992). Burial disturbances in long-lasting shell middens are caused by subsequent interments, digging to retrieve human remains for ritual purposes, other cultural activities (caching material possessions or clearing midden areas for habitation), tree root growth and deadfalls, carnivore and rodent activity, and erosion from sea waves and tides.

Less than 12% of all known midden burials bracket the 13 centuries between the Middle Developmental Stage and the time of European contact; two-thirds of them are from one site. For whatever reason, midden interment seems to have diminished subtantially as a method of corpse disposal in the Late Stage and seems to have entirely ceased by about A.D. 1200–1300, as based on evidence at Greenville (Cybulski, 1992). Outside of that site, the latest well-dated human midden burial is from Little Qualicum River, Strait of Georgia, with an associated radiocarbon date of 730 ± 80 years B.P. (Bernick, 1983).

With 59 individuals, Greenville may represent one of the last examples of midden burial as a significant cultural practice. Twenty-five other Late Stage burials are from 11 middens, an average of 2.3 burials per site, hardly indicative of the large cemeteries known from the Middle Developmental Stage (upwards of 70 to 100 individuals at Prince Rupert Harbour or Strait of Georgia sites). Based on the evidence at Greenville, where two chronological components are known, a possibly ancient coastal ritual of shell-related burial lasted until A.D. 1000, though below-ground interment sporadically continued for another 200 or 300 years. There is a subsequent 500 year-long gap in the archaeological record of Northwest coast mortuary practices, perhaps signifying the introduction of above-ground corpse disposal, a major cultural change that could not have resulted from European contact.

Head Shaping

Intentional head-shape modification is known historically among the central and southern coast tribes of British Columbia. Three different forms may be identified, perhaps corresponding with three regional social networks (Suttles, 1990). In terms of known historic crania, the most northerly network included Heiltsuk, Nuxalk, and Ooweekeno, the next included Kwakwaka'wakw, Kwa-giulth, and Nuu-chahnulth, and the southernmost, Coast Salish. In each network, members may have recognized others of their system by the shape of their heads, cross-cutting language boundaries. Ethnographies indicate, as well, that practicing groups did not include the children born of captive slaves who did not have the social privileges of free men and women.

The antiquity of the practice in the Kwakwaka'wakw, Kwa-giulth, and Nuu-chahnulth region, where heads were intentionally elongated, is unknown. Judging from the lack of evidence at Namu (Curtin, 1984), head shaping in the form of exaggerated roundness may have been a very late innovation on the central coast. For Strait of Georgia, intentional head-shape modification—frontolambdoidal flattening or like forms—was rare or absent before 500 B.C. (Beattie, 1985). After contact, head-shape modification began to decline in some areas in the 1860s (Cybulski, 1978), one type of cultural change brought by European influences.

Labrets and Dental Labret Abrasion

The custom of wearing labrets was widespread among northern Northwest coast groups including Haida, Tsimshian, Nisga'a, and Tlingit of Alaska (Niblack, 1890:256–257). The practice may have included Haisla and Heiltsuk, but did not extend farther south in the historic period. Like head-shape modifications to the south, labrets may have signified membership in a regional northern social network (Suttles, 1990). However, head-shape modification was extended equally to both sexes on the central and southern coasts, but only women wore labrets in the north. This may reflect differences in social organization coupled with a need to symbolize rank and status. The whole area of head deformation was one in which descent was reckoned bilaterally, while Tlingit, Haida, Tsimshian (including Nisga'a), and Haisla shared in a northern regional social system based on matrilineal descent (Suttles, 1990).

Long before contact, labrets were in more widespread use, as shown by excavated specimens (Keddie, 1981) and abrasion facets on teeth (Cybulski, 1974). Besides Prince Rupert Harbour (Cybulski, 1974), dental facets from labrets occur in skeletons from Blue Jackets Creek (Severs, 1974), Namu (Curtin, 1984), Greenville (Cybulski, 1992), and at least three Strait of Georgia sites (Cybulski, 1992).

Tooth wear markers indicate that not all people who wore labrets were buried with them. Most intact interments with labret tooth scars are unaccompanied by the artifacts, suggesting that labrets often were removed from the deceased prior to burial, perhaps to be retained as hereditary family wealth objects (Cybulski, 1992). In Northwest coast societies, status was either ascribed (attained through heredity via kinship) or achieved (attained through acquired wealth) (Matson, 1989). Possibly, those individuals with tooth facets but not labrets in burial were recipients of ascribed status.

The earliest definitive evidence for labret use comes from six skeletons at Blue Jackets Creek and Namu dated between 3000 and 2000 B.C. (Murray, 1981; Curtin, 1984). The abrasion facets were on molars and premolars, indicating that labrets were worn on the sides of the mouth ("lateral" labrets). Both sexes had these facets at Blue Jackets Creek, whereas three Namu skeletons were male. No labrets were found with the skeletons.

The earliest occurrence of dental labret wear is possibly from a site at Pender Canal, Strait of Georgia, radiocarbon dated at 5170 ± 220 years B.P. (Carlson, 1986). In the Pender Canal series, at Crescent Beach in the Strait of Georgia dated to the early Middle Developmental Stage, and at Prince Rupert Harbour and Greenville, all known expressions of abrasion facets were in anterior teeth, indicating that labrets were worn above the chin ("medial" labrets).

Only women (75%) showed labret wear at Greenville, while at Prince Rupert Harbour, 19.6% of males and 5.1% of females were affected. With Blue Jackets Creek, the data indicate that the historic northern pattern began only in the Late Stage. Before then both sexes wore labrets, indicating that north coast social patterns may have changed between Middle and Late Developmental times. Indeed, males clearly dominate the Prince Rupert Harbour picture, in marked contrast to the historically expected pattern at Greenville.

Both south coast series show an almost equal distribution of labret wear for males and females (42.1% and 37.5%, respectively, at Crescent Beach; 8.3% and 10% at Pender Canal). This pattern may reflect a bilateral descent system where either sex might have attained high status positions with equal probability. None of the people with tooth abrasion marks at these sites appear to have been buried with their artifacts, though five others were reported with labrets at Pender Canal (Weeks, 1985, 1986). Assuming the "heirloom" hypothesis, this would suggest only ascribed status for labret-related persons at Crescent Beach and either ascribed or achieved status for such persons at Pender Canal.

Two of 19 Prince Rupert Harbour males with tooth facets and three others lacking appropriate teeth for study had been buried with labrets. The five may represent bearers of achieved status, but the majority of males affected in either way seemingly attained status through descent since the artifacts were not present in burial.

Both dental and artifact evidence suggest that labret use diminished in the south with the onset of head-shape modification, probably ending completely about 2,000 years ago. Why one form of a possible somatic identifier for a regional social system may have supplanted another is unknown. If the two forms are equivalent, it might indicate that the historically known social system in the south was of long-standing duration, possibly 5,000 years. The historically known northern social system would appear to have been a comparatively recent innovation, on the order of 1,300 years.

DEMOGRAPHIC HISTORY

Prior to white contact, 200,000 people may have inhabited the Northwest coast, with about 40% living in British Columbia (Boyd, 1990). Within 100 years, the population had been reduced by about 80%, largely due to the introduction of epidemic-producing diseases, but in some instances more lethal intertribal warfare from the early European introduction of firearms (Duff, 1964). Venereal disease from widespread prostitution and alcohol also took their toll (Duff, 1964).

Inferences about prehistoric mortality are limited by the data at hand. The known cemeteries may not have been burial grounds for village, local, or regional populations, but only for individuals selected by class or lineage. For example, the site at Greenville, representing 700 years of history, was almost completely excavated, but yielded only 59 skeletons. Clearly, many people were buried elsewhere. I have suggested lineage burial in middens to account for hereditary brachydactyly at Prince Rupert Harbour (Cybulski, 1988). There is also reason to suspect selective burial by age group and, to some extent, by sex.

Prehistoric and excavated historic samples contrast in their proportions of infants. Whether arranged by temporal or regional units, infants comprise at best only 11.3% of the prehistoric samples but up to 24.2% in their historic counterparts. Though it is tempting to interpret the difference as a reflection of increased infant mortality in the historic period, the likelihood of infant underrepresentation, perhaps because of different burial customs, prohibits such a conclusion.

Infants are so poorly represented at most shell middens that they may have been relegated to special burial areas in prehistoric times. The prehistoric sites show greater between-group than within-group variability in the proportion of infants when compared with the historic sites (Cybulski, 1992). Some sites at Prince Rupert Harbour and Strait of Georgia seem to have more than their share of infants. For example, burials from two Prince Rupert Harbour sites represent only 11.1% of the total series from that region, but they account for 43.5% of the infants. Two Strait of Georgia sites account for 14% of that region's total, but their samples contain 42.1% of the infants.

Disparate adult sex ratios between the large regional entities of Prince Rupert Harbour and Strait of Georgia may indicate regional differences in sociocultural dynamics. At Prince Rupert Harbour, the ratio of males to females is 1.86:1, while at Strait of Georgia, it is nearly equal at 1.09:1 (Cybulski, 1992). Frequencies of antemortem or perimortem skeletal trauma in the adults also contrast markedly at 49.1% and 17.3%, respectively (48.4% and 14.2% for specifically Middle Stage samples from the two groups), with the majority of Prince Rupert Harbour injuries plausibly attributable to interpersonal violence (Cybulski, 1990, 1992). Warfare may have been frequent at Prince Rupert Harbour, perhaps motivated in part by a desire for women as important resources (Adams, 1973). Captive slaves may have cyclically made up a substantial proportion of the female Prince Rupert Harbour population and, because of their lack of social status, not buried in the same sites as higher ranking residents, largely comprised of men. These circumstances may not have existed at prehistoric Strait of Georgia.

Regional differences in the Middle Developmental Stage, as well as small samples in the Early and Late Stages, make it difficult to study accurately possible coastwide trends in skeletal trauma and warfare-related injuries through time. The available adult samples show an increase in trauma from 28.1% in the Early Stage (57 observable persons) through 44.9% in the Late Stage (58 persons), but a frequency of only 21.1% in the historic period (218 persons). Injuries due to interpersonal violence remain stable through the Early and Middle Stages (21% and 21.2%), increase in the Late Stage (27.6%), and are significantly lower in the historic period (12.8%). As in the prehistoric Strait of Georgia case, the sex ratio of adults from excavated historic sites is nearly equal. The Prince Rupert Harbour region is not represented in the historic sample.

HEALTH AND DISEASE

Smallpox, measles, influenza, whooping cough, scarlet fever, and malaria reached epidemic proportions among Northwest coast Indians soon after contact (Boyd, 1990). We know little of such diseases in the prehistoric period since acute infections rarely involve bone. Chronic bone infections likely due to tuberculosis and treponematosis appear in the historic record, as do malignant bone tumors (Cybulski, 1990). Tuberculous spinal lesions are known from Nuu-chah-nulth and Coast Salish skeletons, treponemal cranial vault lesions from Haida, Ooweekeno, and Coast Salish skeletons, and malignant bone tumors from Haida and Coast Salish remains. Of these three disease groups, only treponemal infections appear in the prehistoric record, in Middle Developmental Stage remains at both Strait of Georgia and Prince Rupert Harbour where both cranial vault and long bone lesions are known (Cybulski, 1990, 1991; Skinner, 1991). An 1860s writer living among the Nuu-chah-nulth concluded that "syphilis," which was quite common at that time, was probably indigenous "as the natives have herbs which they use in curing this disease." (Sproat, 1987:168).

Cribra orbitalia (porotic hyperostosis), a possible indicator of iron-deficiency anemia, is at equal frequencies of 13–14% in coastwide prehistoric and historic samples, but regionally variable (Cybulski, 1990). This skeletal lesion may reflect a population's successful adaptation to the pathogen load of its local environment, including all fungi, viruses, bacteria, and parasites to which a human group is subjected (Stuart-Macadam, 1992). The extent of the load is dependent on "climate, geography, topography, population size and density, hygiene, food resources, seasonality, customs, and subsistence patterns" (Stuart-Macadam, 1992:42). It is difficult to identify any one factor to account for the occurrence of porotic hyperostosis in any one population. Climate, geography, topography, and, probably, resources, seasonality, customs, and subsistence patterns would have been similar in most coastal British Columbia groups, leaving variations in population size and density, and, perhaps, sanitation, to account for local or regional variations in the frequency of cribra orbitalia.

Stature, a possible indicator of nutritional health, seemingly changed little until well into the twentieth century, with skeletally estimated averages of 162.4 cm and 163.3 cm in Early Stage and historic period men (Cybulski, 1990). Native men measured by Franz Boas and coworkers late in the nineteenth century averaged 164.2 cm (Cybulski, 1990), and Nuu-chah-nulth men measured in the 1960s averaged 170.4 cm (Birkbeck et al., 1971).

Dental caries show significant differences in frequencies of occurrence between the prehistoric and historic skeletal samples, but with per tooth figures of only 0.3% and 0.8% (samples in excess of 6,000 teeth). They are far lower than those reported for native and non-native populations elsewhere in North America through the nineteenth century (Cybulski, 1990, 1992). Anecdotal references suggest that it was not until the early twentieth century that tooth decay became a real problem among Northwest coast Indians (Large, 1968). For example, studies on native reserves in the 1920s reported per tooth caries frequencies of 36.9% and 48.5% owing to a reliance on non-native foods rich in carbohydrates (Price, 1934). Traditional native diets consisted principally of fish, supplemented by sea and land mammal meats and some plant foods (e.g., McNeary, 1976).

Periapical abscesses appear to have been more of a health problem than tooth decay in the prehistoric and contact periods with per tooth socket frequencies of 9.1% and 3%, respectively. Severe occlusal attrition leading to open tooth pulp chambers probably resulted from traditional food preparation techniques that introduced grit to the food (Cybulski, 1990).

SYNOPSIS

Table 1 is a synoptic representation of the skeletal and mortuary-derived information of this paper. The data suggest few changes were brought to Northwest coast Indians during the first 100 years of European contact. Dental health began to change, but only slightly, as, perhaps, did stature. The only substantial change may have involved selective burial practices with respect to infants. The lack of evidence for tuberculosis and malignant bone tumors in the prehistoric period may be a function of age-related sampling and poor bone preservation, especially when these conditions produce osteolytic lesions that weaken bone. Skeletal tuberculosis is often a disease of the young, while malignancies are often associated with older people. Both conditions have been chronicled in other precontact American Indian remains (Steinbock, 1976).

Substantial changes with respect to culture and, possibly, social organization appear to have occurred on the coast of British Columbia in prehistoric times as reflected in methods of corpse disposal, head-shape modification, and the use of labrets.

Considering the clear record for altered customs and physical status in distant time, it is interesting that comparable changes cannot be picked up by mortuary and skeletal data for the early contact period when native groups elsewhere in the Americas were so greatly affected by the arrival of Old World peoples. In fact, the clearest indications of historic period change on the Northwest coast pertain to a very recent date, only a century ago. In part, this may reflect a tenacious adaptation of Northwest coast peoples to an environment long considered barren and hostile by potential European settlers. The early contact period was dominated by the maritime fur trade, and Northwest coast Indians were long-established, skilled traders on an intertribal basis. Actual white colonization and settlement did not occur until the 1830s and 1840s, and only in a few major mainland centers considered suitable for the introduction of European agriculture (Cole and Darling, 1990). These findings indicate, then, that biological effects of contact did not occur until sustained, long-term contact became manifest in the middle to late nineteenth century on the Northwest coast.

ACKNOWLEDGMENTS

Many people contributed information on which this paper is based, though none are responsible for my possible misinterpretations of the data. In particular, I thank the people of Hesquiat, Rivers Inlet, and Greenville for allowing me access to their ances-

TABLE 1. Synopsis of Skeletal and Mortuary-Derived Issues on the British Columbia Coast Through Time

Issue	Early (3500–1500 B.C.)	Middle (1500 B.C.–A.D. 500)	Late (A.D. 500–1774)	Historic (A.D. 1774–1874)
Dated human remains	77	584	84	593
Corpse disposal	Below ground —	— through —	— A.D. 1250	Above ground
Head shaping				
North coast	No	No	No	No
Central coast	No	No?	No?	Yes
Strait of Georgia	No	500 B.C. —	— through —	—1860s
Labret use				
North coast	Both sexes	Males > females	Females	Females
Central coast	Males	No data	No data	Females?
Strait of Georgia	Males = females —	—to 500 B.C.	No	No
Selective burial				
Infants	Special burial —	—areas through—	— A.D. 1250?	No
Females	Insufficient data	Yes (north)/No (south)	Insufficient data	No
Adult trauma (%)	28.1	48.4 (north)/14.2 (south)	44.9	21.1
Violence (%)	21.1	32.4 (north)/6.2 (south)	27.6	12.8
Treponematosis	No evidence	Present	No evidence	Present
Tuberculosis	No evidence	No evidence	No evidence	Present
Malignant tumors	No evidence	No evidence	No evidence	Present
Cribra orbitalia (%)	←	13.8	→	13.1
Male stature (cm)	162.4	162.9	162.6	163.3/164.2
Dental caries (%)	←	0.3	→	0.8
Dental abscesses (%)	←	9.1	→	3.0

tral burial grounds, and the people of Metlakatla, Waglisla, Bella Coola, Nanaimo, Cowichan, and Semiahmoo for entrusting their ancestral skeletal remains to my analysis. Colleagues Joanne Curtin and Owen Beattie kindly shared data. I am grateful to archaeologists Roy Carlson, James Haggarty, George MacDonald, R.G. Matson, Elaine Moore, James Pike, Bjorn Simonsen, Patricia Sutherland, and the late Carl Borden for allowing me access to collections in their care, as well as to staff of the Royal British Columbia Museum, the Field Museum of Natural History, and the American Museum of Natural History. To all of the above, and to Philip Hobler, Alan McMillan, and Mark Skinner, I am also grateful for their support, stimulating thoughts, and encouragement over the years.

REFERENCES CITED

Adams JW (1973) The Gitksan Potlatch: Population Flux, Resource Ownership and Reciprocity. Toronto: Holt, Rinehart and Winston.

Adams JW (1981) Recent ethnology of the Northwest coast. Annual Review of Anthropology *10*: 361–392.

Beattie OB (1985) A note on early cranial studies from the Gulf of Georgia region: Long-heads, broad-heads, and the myth of migration. BC Studies *66*:28–36.

Bernick K (1983) A Site Catchment Analysis of the Little Qualicum River Site, DiSc 1: A Wet Site on the East Coast of Vancouver Island, B.C. National Museum of Man Mercury Series, Archaeological Survey of Canada Paper 118.

Birkbeck JA, Lee M, Myers GS, and Alfred BM (1971) Nutritional status of British Columbia Indians. II: Anthropometric measurements, physical and dental examinations at Ahousat and Anaham. Canadian Journal of Public Health *62*:403–414.

Boyd RT (1990) Demographic history, 1774–1874. In W Suttles (ed.): Handbook of North American Indians, vol. 7, Northwest Coast. Washington, DC: Smithsonian Institution Press, pp. 135–148.

Burley DV (1989) Senewelets: Culture History of the Nanaimo Coast Salish and the False Narrows Midden. Royal British Columbia Museum Memoir 2.

Carlson RL (1986) The 1985 Excavations at the Canal Site (DeRt 1 and DeRt 2). Manuscript on file (Permit 1985-10), Archaeology Branch, Province of British Columbia, Victoria.

Cole D, and Darling D (1990) History of the early period. In W Suttles (ed.): Handbook of North American Indians, vol. 7, Northwest Coast. Washington, DC: Smithsonian Institution Press, pp. 119–134.

Collison WH (1981) In the Wake of the War Canoe. Based on the edition originally published in 1915, edited and annotated by Charles Lillard. Victoria: Sono Nis Press.

Curtin AJ (1984) Human Skeletal Remains from Namu (ElSx 1): A Descriptive Analysis. M.A. thesis, Department of Archaeology, Simon Fraser University, Burnaby.

Cybulski JS (1974) Tooth wear and material culture: Precontact patterns in the Tsimshian area, British Columbia. Syesis *7*:31–35.

Cybulski JS (1978) An Earlier Population of Hesquiat Harbour, British Columbia: A Contribution to Nootkan Osteology and Physical Anthropology. Royal British Columbia Museum, Cultural Recovery Papers 1.

Cybulski JS (1988) Brachydactyly, a possible inherited anomaly at prehistoric Prince Rupert Harbour. American Journal of Physical Anthropology *76*:363–376.

Cybulski JS (1990) Human biology. In W Suttles (ed.): Handbook of North American Indians, vol. 7, Northwest Coast. Washington, DC: Smithsonian Institution Press, pp. 52–59.

Cybulski JS (1991) Human Remains from Duke Point, British Columbia, and Probable Evidence for Pre-Columbian Treponematosis. Manuscript on file (No. 3454), Library Documents Section (Archaeology), Canadian Museum of Civilization, Hull.

Cybulski JS (1992) A Greenville Burial Ground; Human Remains and Mortuary Elements in British Columbia Coast Prehistory. Hull: Canadian Museum of Civilization.

Duff W (1964) The Indian History of British Columbia, vol. 1, The Impact of the White Man. Anthropology in British Columbia Memoir 5.

Fladmark KR (1986) British Columbia Prehistory. Ottawa: National Museums of Canada.

Keddie GR (1981) The use and distribution of labrets on the North Pacific rim. Syesis *14*: 59–80.

Large RG (1968) Drums and Scalpel; From Native Healers to Physicians on the North Pacific Coast. Vancouver: Mitchell Press Limited.

MacDonald GF, and Inglis RI (1981) An overview of the north coast prehistory project (1966–1980). BC Studies *48*:37–63.

Matson RG (1989) The Locarno Beach phase and the origins of the Northwest Coast ethnographic pattern. Reprint proceedings of the Circum-Pacific Prehistory Conference, Seattle, vol. III, part 2, pp. 1–44.

McNeary SA (1976) Where Fire Came Down: Social and Economic Life of the Niska. Ph.D. dissertation, Department of Anthropology, Bryn Mawr College, Bryn Mawr.

Mitchell DH (1971) Archaeology of the Gulf of Georgia Area, A Natural Region and Its Culture Types. Syesis 4 (supplement 1):1–228.

Moser C (1926) Reminiscences of the West Coast of Vancouver Island. Victoria: Acme Press Ltd.

Murray JS (1981) Prehistoric skeletons from Blue Jackets Creek (FlUa 4), Queen Charlotte Islands, British Columbia. In JS Cybulski (ed.): Contributions to Physical Anthropology, 1978–1980. National Museum of Man Mercury Series, Archaeological Survey of Canada Paper 106, pp. 127–168.

Niblack AP (1890) The Coast Indians of Southern Alaska and Northern British Columbia. Report of the U.S. National Museum for 1888, pp. 225–386.

Price WA (1934) Why dental caries with modern civilizations? VIII. Field studies of modernized Indians in twenty communities of the Canadian and Alaskan Pacific coast. Dental Digest *40*:81–84.

Severs PDS (1974) Archaeological investigations at Blue Jackets Creek, FlUa-4, Queen Charlotte Islands, British Columbia, 1973. Canadian Archaeological Association Bulletin 6:165–205.

Skinner MF (1989) Salvage recovery from an ossuary cave and rock crevice burial complex on Gabriola Island, (DgRw 199). Canadian Association for Physical Anthropology Newsletter, Spring 1989, p. 19.

Skinner MF (1991) Osseous Treponematosis of the Americas: A Review with New Evidence from Gabriola Island (DgRw 199), B.C., Canada. Manuscript on file (Permit 1988-1), Archaeology Branch, Province of British Columbia, Victoria.

Sproat GM (1987) The Nootka; Scenes and Studies of Savage Life. Based on the edition originally published in 1868, edited and annotated by Charles Lillard. Victoria: Sono Nis Press.

Steinbock RT (1976) Paleopathological Diagnosis and Interpretation. Bone Diseases in Ancient Human Populations. Springfield: Charles C. Thomas.

Stuart-Macadam, P (1992) Porotic hyperostosis: A new perspective. American Journal of Physical Anthropology *87*:39–47.

Suttles W (1990) Introduction. In W Suttles (ed.): Handbook of North American Indians, vol. 7, Northwest Coast. Washington, D.C.: Smithsonian Institution Press, pp. 1–15.

Weeks SM (1985) Skeletal remains from DeRt 2. In RL Carlson: The 1984 Excavations at the Canal Site (DeRt 1 and DeRt 2). Manuscript on file (Permit 1984-13), Archaeology Branch, Province of British Columbia, Victoria, pp. 78–118.

Weeks SM (1986) Skeletal remains from DeRt 1 and DeRt 2. In RL Carlson: The 1985 Excavations at the Canal Site (DeRt 1 and DeRt 2). Manuscript on file (Permit 1985-10), Archaeology Branch, Province of British Columbia, Victoria, pp. 94–116.

Chapter 8

Historic Epidemics of the American Pueblos

Ann M. Palkovich

Department of Sociology and Anthropology, George Mason University, Fairfax, VA 22030

Persistence and transformation characterize the early historic period of southwestern Pueblo groups. In contrast to many other Native American groups, Old World diseases did not extinguish Pueblo populations, although these societies were changed forever. Despite years of study, scholarly opinion is still divided over the timing of initial Old World disease introduction into Pueblo populations, the spatial pattern of epidemics among Pueblos, and the role of trade centers in the spread of these diseases. Historic documents, although far from a perfect source of information, can be used to address controversies over the extent and pattern of epidemic disease loss among Pueblo populations.

Dobyns (1983:14), a leading proponent of large precontact population size and its rapid, dramatic decline, has argued forcibly that any discontinuity in the archaeological or historic record is sufficient evidence to invoke immediate and massive epidemic mortality as the primary cause. For Dobyns (1983), the Americas in A.D. 1492 were thickly populated, suggesting a population several times greater than most current estimates. Shortly after the Europeans arrived, large numbers of Indians throughout the New World rapidly succumbed to newly introduced deadly diseases, such as smallpox. These diseases often spread beyond intial points of introduction to affect uniformly most, and likely all, populations distributed widely across large segments of the Americas. The broad transmission of contagious diseases was enhanced by regular contacts maintained among peoples within and between cultural areas, such as the American Southwest. For example, Dobyns (1983: 14–15) claims the Pueblos were part of a New World smallpox pandemic in 1520–1524 and that a catastrophic depopulation of the southwestern Pueblos took place at the same time as in northwestern Mexico. Dobyns concludes that a massive Pueblo population loss of nearly 75% prior to established Spanish contact in 1539 left only a remnant population that was vulnerable to continued epidemic outbreaks of several new diseases, further reducing the size of the Pueblo populations. Both the documentary and archaeological evidence suggest otherwise.

EARLY HISTORIC POPULATION

The year 1539 serves as a watershed for the American Southwest. It is widely agreed

In the Wake of Contact: Biological Responses to Conquest, pages 87–95

that the earliest Spanish contact with Pueblo settlements dates to the De Niza *entrada* of 1539; it inspired the Coronado *entrada* of 1540 (Hammond and Rey, 1940:254–256). After a lapse of some 40 years, other Spanish expeditions to the Pueblos followed in 1581, 1582, and 1591 (Schroeder, 1979). In 1598, Spanish efforts to colonize the area were initiated by Oñate (Simmons, 1979).

Various scholars (Simmons, 1979; Upham, 1982, 1992; Reff, 1992) estimate that the combined population for the Eastern and Western Pueblos was approximately 100,000 to 130,000 at the time of contact in 1540. These estimates essentially double Oñate's 1598 population figure of 60,000 (Bolton, 1916). It also places population loss for the century following 1540 at 60% to 70%, using Prada's 1638 figure of 40,000 (Hackett, 1937).

Refining such figures for the timing and extent of Pueblo population loss is difficult. Records indicate 60,000 Puebloans were baptized between 1600 and 1638 (Hackett, 1937). Oñate's and Prada's population figures suggest a 30% loss in 40 years. However, if major epidemics did not reach Pueblo settlements until 1606, the potential reduction in just 30 years could have represented 60% of the estimated 100,000 individuals.

EPIDEMICS AND INITIAL CONTACT

Contact between the Pueblos of the American Southwest and groups in northern Mexico are essential conditions for frequent precontact transmission of epidemic diseases northward. Reff (1991, 1992) argues that large portions of northern Mexico were sparsely populated by small, mobile bands in the sixteenth century. Long-distance trade between Mesoamerica and the Southwest had collapsed by the mid-fourteenth century (Reff, 1991:102–103). In addition, hostilities between Chichimec and Mexica groups limited their contacts, thus blocking the spread of smallpox northward (Reff, 1991:103). Sixteenth-century documents for northern Mexico do not report epidemics or excessive deaths. Reff concludes that the smallpox epidemic of 1518–1525 did not extend beyond central Mexico. It appears that the systematic exploration and ultimate colonization of the American Pueblos became feasible only after the development of a transportation network in northern Mexico built to promote mining. Judging from Reff's (1991:149) analysis, smallpox and measles did not reach the American Pueblos until the broad-scale epidemic of 1606–1607.

Reff's (1991) study is consistent with archaeological evidence for the American Southwest. Spanish disruption of the Pueblos must be placed in the context of successive regional population movements usually seen as responses to water shortages (e.g., Upham, 1984; Cordell and Gumerman, 1989). By 1500, the Pueblo area, once dotted with settlements ranging from small villages to family dwellings, featured large and often multistoried room-block complexes surrounding central plazas.

Most late prehistoric communities appear to have had "weak social hierarchies" and participated in local exchange networks (Cordell and Gumerman, 1989:13). Such social organization has been characterized by Johnson (1989:378) as sequential social hierarchies featuring "organization of consensus among basically egalitarian aggregates of increasing inclusiveness," such as families, residence groups, and clans. Late prehistoric Pueblo settlements appear to have been comprised of many family groupings in large room blocks that aggregated in response to demands for increased cooperation and greater labor requirements to maintain an agricultural subsistence base. Although high-mortality historic epidemics would have disrupted these social units, it is unlikely that even major population loss would shatter such basic forms of readily regenerated social integration. During the historic period, residence groups or clans simply recombined when faced with the challenges of crop failure and population loss.

Some southwestern archaeologists (e.g., Mathien and McGuire, 1986), however, have argued for more complex social systems. These include "statelets" (Riley, 1986, 1987) or hierarchically organized, economically and politically linked villages dependent on long-distance trade of elite items (Upham, 1982). Social and political disintegration are more likely for finely balanced hierarchically organized societies faced with great depopulation, and would likely have resulted in an abandonment of large villages for small settlements.

Opposing views of social organization and trade in the late prehistoric period have important implications for the expected impact and spread of human-transmitted diseases such as smallpox and measles in early historic times. Disruptions of existing social orders have very different effects if distinct, independently mobile social modules are challenged rather than highly interdependent, politically integrated villages.

Dobyns (1983:303) interprets Castaneda's 1540 reports of abandoned Pueblos and partially filled settlements as evidence of precontact depopulation. However, population movement, resettlement, and settlement abandonment are entirely consistent with repeated patterns of coalescence and dispersal of prehistoric Pueblo groups over many centuries (e.g., Cordell and Gumerman, 1989). For the American Southwest, citing Pueblos abandoned in the late prehistoric period as primary evidence for massive precontact population loss does not make a compelling argument.

PATTERNS OF EPIDEMICS

Accounts of epidemics during the eighteenth and nineteenth centuries are incomplete and ambiguous (Table 1). Although major outbreaks, such as those experienced in 1638 and 1781, appear in many records, information about disease prevalence in individual Pueblos is, at best, sketchy. Even the major smallpox epidemic of 1781, said to have killed 5,025 of the 9,104 Pueblo inhabitants counted in that year (Bancroft, 1888), is inconsistently reported. Franciscans at only Sandia, Santo Domingo, and Santa Clara made note of the epidemic and resulting deaths. These communities were only a few of the Pueblos occupied at that time. Other problems are apparent in the records, such as the difficulty of pinpointing specific diseases in ambiguous accounts (Upham, 1986, 1987; Reff, 1987) and the reporting of an epidemic in one account (e.g., Table 1, Cochiti in 1896) contradicted by a second report of a prospering community (Starr, 1899).

One means of assessing the pattern of epidemic disease among the Pueblos is to link reports of epidemic outbreaks with population estimates and burial records. Such information is currently available for San Juan and Pecos Pueblos.

Aberle et al. (1940) identify nine major epidemic episodes from daily records of burials for San Juan (Table 2). From 1726 to 1849, they note marked fluctuations in the number of deaths, with epidemics apparently doubling or tripling the usual number of deaths that typically occurred in a single year (Aberle et al., 1940:171). This pattern suggests that mortality was significantly increased by disease outbreaks, but populations were not necessarily devastated each time a disease hit the Pueblos. In addition, Aberle et al. (1940:169, 176) note that epidemics did not necessarily strike San Juan Puebloans and local Spanish residents at the same time, and that greater mortality was evident among the Spanish population in the area than among Indians residing at San Juan. Apparently, by the early eighteenth century the San Juan population did not experience any special susceptibility to epidemics.

TABLE 1. Epidemics for Historic New Mexican Pueblos[a]

Year	Pueblo	Epidemic
1638	Widespread	Cocolitzli (smallpox) (Hackett, 1937)
1640	Widespread	"*Peste*," 3,000 deaths (Simmons, 1966)
1719	Nambe	Smallpox, 24 adults and 21 children die
1728	Jemez	Pestilence, 108 die (Hodge, 1910)
	Galisteo	Many deaths, Dec.
	Acoma	Measles, Dec.–Jan.
	Pecos	Measles, Dec.–Jan. (Kessell, 1979:377)
1733	Santa Ana	Smallpox, April–Aug.
	Jemez	Smallpox, June–July
1734	Zuni	Smallpox, 230 deaths
1737	Santa Cruz	Unknown, many children die, 67 deaths
1738	Pecos	Smallpox, Jan.–March
1748	Pecos	Unknown, August
1770	Santa Clara	Unknown, many children die, May–July
1772	Laguna	"*La Peste*," 44 adults, 21 children die, Jan. 4–March 31
1781	Sandia	Smallpox, Jan.–Oct. (Simmons, 1966)
	Santo Domingo	Smallpox, 230 deaths, Feb. 4–March 9, other smaller epidemics
1781–1782	Santa Clara	Smallpox, Dec.–March
1782	San Juan	Smallpox, 500 deaths with Santa Clara in two months (Simmons, 1966)
1782–1794	Galisteo	Smallpox (Hodge, 1910)
1799–1800	Santa Cruz	Smallpox, Nov.–March
1800	Santa Clara	Smallpox, Jan.
1804	Santa Cruz	Unknown, increased deaths, May–June
1805	Santa Clara	Measles or cholera? April–May
	Santa Clara	Measles
	Cochiti	Measles, April
1816	Santo Domingo	Smallpox, 78 deaths, Sept. 4–19
	San Juan	Smallpox, Spring
	Picuris	Unknown, increased deaths
1817	Santo Domingo	Unknown, 22 deaths, Jan. 11–21
1840	San Juan	Fever in adults, smallpox in children, Spring
1853	Zuni	Smallpox (Hodge et al., 1945)
1883	Taos	Smallpox (Sanchez, BIA)
1887	Isleta	Measles, 30–35 deaths (Sanchez, BIA)
1889	Jemez	Smallpox, diphtheria (Williams, BIA)
	Cochiti	Smallpox, diphtheria (McClure, BIA)
	Isleta	Smallpox, diphtheria (McClure, BIA)
	San Felipe	Smallpox, diphtheria (McClure, BIA)
	Santa Ana	Smallpox, diphtheria (McClure, BIA)
	Zia	Smallpox, diphtheria (McClure, BIA)
1891	Acoma	Smallpox, diphtheria (Segura, BIA)
1894	Cochiti	"Mountain fever", many die, 26 die in one week (Starr, 1899)
1895	Santo Domingo	Spinal meningitis or spotted fever, 42 deaths (Bullis, BIA)
1896	Cochiti	Malaria (Dumarest, 1919; contra Starr, 1899)
1898–1899	Zuni	Unknown, killed about 250 (Walpole, BIA)
1899	Cochiti	Malaria, 30 deaths (Walpole, BIA)
1900	Cochiti	Malaria, 20 deaths (Walpole, BIA)
	Isleta	Measles, 20 deaths (Walpole, BIA)

[a]Data from Chavez (1957) except where noted. For specific Bureau of Indian Affairs (BIA) Annual Reports, U.S. Department of the Interior, citations see Palkovich (1985).

TABLE 2. Major Epidemics, San Juan Pueblo[a]

Year	Epidemic
1781	Smallpox, 95 deaths
1788–1789	Smallpox, Dec. 11–Feb. 26
1800	Smallpox, Jan. 8–Feb. 19
1816	Unknown, March 12–June 11; July 30–Dec. 24
1826–1827	Unknown, Jan. 8–April 16
1846–1847	Unknown, Nov. 20–Jan. 22
1855–1859	Unknown, deaths somewhat evenly distributed over five-year period
1890–1891	Unknown, Sept. 11–Jan. 29
1899	Unknown, Jan. 29–June 4

[a]Data from Aberle et al. (1940:167–69).

Kessell (1979:378) identifies 10 major epidemics from "the bunching of deaths in the Pecos burial books" (Table 3). With major epidemics nearly every decade and others in intervening years, he concludes epidemics caused considerably more deaths at Pecos than the well-known attacks of Plains Indians (Kessell, 1979:378).

SPOTTY EPIDEMIC PATTERN

This admittedly scanty information suggests a mosaic pattern of major epidemics among the Pueblos. Smallpox was reported at both Pecos and San Juan Pueblos in 1781 and 1800. Smallpox again struck Pecos in 1826 and may have been present at San Juan in that year as well. Four more major epidemics hit San Juan prior to 1900; however, Pecos was abandoned in 1838. Aberle et al. (1940) note no significant cluster of San Juan deaths between 1726 and 1781, indicating no major epidemic outbreaks, yet six major epidemics are noted for Pecos by Kessell (1979). Clearly, epidemics did not uniformly affect the Pueblos, as implied by Dobyns (1983:13). Scholars like Dobyns (1983) cannot assume that a report of a new deadly disease in a particular area necessarily means that all local communities in the region—Spanish and Pueblo alike—experienced the illness and all those struck by the illness were equally affected.

TABLE 3. Major Epidemics, Pecos Pueblo[a]

Year	Epidemic
1696	Fever? May 1, 10 burials (deaths attributed to cough [1], pain/cough [4], fever [5])
1704	Unknown
1728–1729	Measles
1738	Smallpox, in 18 weeks 26 young children die
1748	Unknown, late summer, at least 15 children and three men die; 13 men die in Jan. Comanche assault
1759	Unknown
1780–1781	Smallpox, burial books for Pecos missing
1800	Smallpox
1816	Smallpox, 18 adults die in Dec. alone
1826	Unknown, seven children die in two days among a total population of 40

[a]Data from Kessell (1979:378).

For both Pecos and San Juan, reported epidemics or years of excessive deaths do not consistently correspond with population loss as reflected in estimates of Pueblo occupants (Palkovich, 1985:414, 419). Population figures for both Pueblos reflect increases as well as declines in the rate of change in the population. Demographic trends for irregular time intervals are most accurately represented by a "coefficient of population movement." This measure provides the percent change in population growth or decline averaged over the number of years between the two population estimates, thus estimating the relative change in population size. Dramatic reversals in demographic trends between population declines and increases are evident for all historic Pueblos (Palkovich, 1985:413–417). No consistent rate of depopulation can be demonstrated for the Pueblos using solely historic population estimates. This pattern contrasts with the steady rates of decline Cook and Borah (1971), among others, document for historic Mexican populations. Apparently, while epidemics were reducing the overall Eastern Pueblos population, this pattern is not clearly reflected in the population figures for individual Pueblos.

Inconsistent reporting of population figures, varying enumeration criteria, and missing records introduce some ambiguities into these demographic estimates (Palkovich, 1985). Migration of individuals among various Pueblos is noted throughout the eighteenth and nineteenth centuries. However, Dobyns's (1983:314) contention that epidemics were the universal, primary cause of population migration and settlement consolidation among historic Native American groups, including the Pueblos, greatly overstates the case. Intergroup hostilities, famine, migration, and political realignments also played major roles in historic demographic and cultural transformations (e.g., Kessell, 1979). For example, on January 21, 1748, 13 men died in an assault on Pecos by the Comanches; late that summer, at least 15 children and three men died in an epidemic (Kessell, 1979:378). Death records, particularly for Pecos, are punctuated with such reports of deaths from both violence and disease. Further investigations of the burial books for each Pueblo are needed to chart the spatial and temporal epidemiologic patterns of both major and minor disease outbreaks throughout the eighteenth and nineteenth centuries.

PECOS AS A TRADE CENTER

Dobyns (1992:216–220) cites Pecos as a Native American trade center where frequent intergroup exchanges facilitated the spread of epidemic diseases. As a result, these centers experienced excessive mortality that ultimately led to their early abandonment. Dobyns assumes that catastrophic precontact population loss in a densely populated settlement was necessary in order to account for Pecos's abandonment. He implies that famine and warfare at Pecos resulted directly from the "killing years" of epidemics in the late eighteenth and early nineteenth centuries.

Characterizing Pecos as a community driven to extinction solely by epidemics is contrary to this village's documented history. Kessell (1979) attributes the declining population of Pecos to four intimately linked factors; namely, disease, famine, emigration, and conflict. Historic estimates give Pecos's population as approximately 2,000 individuals in the seventeenth century (Kessell, 1979:489). Throughout the historic period, Pecos's population was in continual decline, dwindling from about 1,000 individuals in 1700 to 449 by 1750 (Kessell, 1979:335). In 1776, 269 residents are reported; after 1779, the population permanently fell below 200. The final 18 survivors abandoned the settlement in 1838.

Pecos was of strategic importance for the Spanish throughout the seventeenth century (Kessell, 1979:369). However, the early decades of the eighteenth century were a time of significant decline for Pecos. Famines resulted from crop failure, an ever-present possibility in a marginal environment, and the abandonment of needed fields that were too dangerous to tend because of Comanche raids (Kessell, 1979:347). Famine undoubtedly exacerbated the impact of disease outbreaks, and hard times resulted in a steady emigration from Pecos. By the mid-1700s, Pecos no longer had a resident friar, and the Pueblo's status was reduced to that of a *visita* (Kessell, 1979:349).

As the easternmost Pueblo in the historic period, Pecos served as the point of exchanges between Pueblo and Plains groups. Its location also made it uniquely vulnerable to hostilities, particularly from the Apache and Comanche (Spielmann, 1983, 1989). Sporadic trade fairs were held into the nineteenth century despite steadily increasing hostilities from the Comanches. Pecos's decline as a center of commerce was driven by the displacement of Plains groups leading to frontier hostilities with the Pueblo for nearly a century prior to its abandonment (Kessell, 1979:371).

Excessive deaths attributed to epidemics at Pecos between 1696 and 1780, which were not evident at San Juan, suggest Pecos may have experienced more exposure to introduced Old World diseases than other communities. However, the distance and relative isolation of Pecos from the rest of the Rio Grande Pueblos, along with the lack of correspondence of disease outbreaks elsewhere, suggest Pecos did not serve as a focal point of epidemics that spread widely.

CONCLUSION

Survival with change aptly describes the contact period for southwestern Pueblo societies, despite novel pathogens, political subjugation, and heightened hostilities. It can be argued that the Pueblos persisted because the Spanish were dependent on these Native Americans for their basic survival in this meager land. But over time, epidemic diseases certainly diminished their numbers. We will never know precisely when smallpox and other Old World diseases first descended on the Pueblos. Currently, the best evidence suggests that smallpox arrived no earlier than 1540, and it can be argued that smallpox, measles, and other great killers first struck the Pueblos no earlier than 1606. Clearly, introduced diseases were taking their toll by the mid-1600s when Spanish colonization was firmly entrenched. Population losses of up to 60–70% between 1606 and 1638 seem likely.

Epidemic outbreaks as a uniform, sweeping scourge are not supported by the currently available historic documents, despite their prominence in the work of some scholars. Census and burial figures show that epidemics occurred sporadically among Pueblos, and communities experienced different patterns of major and minor diseases. Historic records for just two Pueblos suggest that sweeping generalizations about the timing and magnitude of epidemic-caused mortality are not justified (e.g., Dobyns, 1983, 1992).

Similarly, the decline of Pecos as a trade center triggered by excessive mortality from infectious disease misrepresents the known history of this settlement. Epidemics no doubt caused many deaths, but they did not play an exclusive role in Pecos's abandonment. Famine and warfare, present among the Pueblos far back into prehistoric times, were also important. The post-1492 exploration and colonization of North America exacted a terrible toll. Nevertheless, our ability to reconstruct the impact of Old World diseases in the Southwest depends, in part, on the extent to which patterns of persistence, transformation, and change can be teased out of historic documents. To date, only a portion of the story has been deciphered. Essential information about the patterns of population loss for each Pueblo remain to be evaluated in the Franciscan burial books. Taken individually, these data, like archaeological and skeletal evidence, leave us with many questions and ambiguities. However, carefully constructed models can be used to make sense of incomplete data. The epidemiological and demographic picture of each Native American group or region requires as detailed a reconstruction as possible before the broader picture of population loss for North American as a whole will emerge.

REFERENCES CITED

Aberle SD, Watkins JH, and Pitney EH (1940) The vital history of San Juan Pueblo. Human Biology *12*:141–187.

Bancroft HH (1888) The Works of H.H. Bancroft. San Francisco: The History Co.

Bolton HE (1916) Spanish Exploration in the Southwest, 1542–1700. New York: Charles Scribner's Sons.

Chavez FA (1957) Archives of the Archdiocese of Santa Fe. Bibliographic Series, vol. 3. Washington, DC: Academy of American Franciscan History.

Cook S, and Borah W (1971) Essays in Population History: Mexico and the Caribbean, vol. 1. Berkeley: University of California Press.

Cordell L, and Gumerman G (eds.) (1989) Dynamics of Southwestern Prehistory. Washington, DC: Smithsonian Institution Press.

Dobyns HF (1983) Their Number Become Thinned. Knoxville: University of Tennessee Press.

Dobyns HF (1992) Native American trade centers as contagious disease foci. In JW Verano and DH Ubelaker (eds.): Disease and Demography in the Americas. Washington, DC: Smithsonian Institution Press, pp. 215–222.

Dumarest N (1919) Notes on Cochiti, New Mexico. Memoirs of the American Anthropological Association 6(3). Lancaster, Pennsylvania.

Hackett C (ed.) (1937) Historical Documents Relating to New Mexico, Nueva Vizcaya and Approaches Thereto, to 1773, vol. 3. Papers of the Division of Historical Research, Publication 330. Washington, DC: Carnegie Institution.

Hammond G, and Rey A (eds.) (1940) Narrative of the Coronado Expedition, 1540–1542. Albuquerque: University of New Mexico Press.

Hodge FW (ed.) (1910) Handbook of American Indians North of Mexico. Bureau of American Ethnology, Bulletin 30.

Hodge FW, Hammond G, and Rey A (eds. and trans.) (1945) Fray Alonso de Benevides's Revised Memorial of 1634. Albuquerque: University of New Mexico Press.

Johnson G (1989) Dynamics of Southwestern prehistory: Far outside—looking in. In L Cordell and G Gumerman (eds.): Dynamics of Southwestern Prehistory. Washington, DC: Smithsonian Institution Press, pp. 371–390.

Kessell JL (1979) Kiva, Cross and Crown: The Pecos Indians and New Mexico, 1540–1840. Washington, DC: National Park Service, United States Department of the Interior.

Mathien F, and McGuire R (eds.) (1986) Ripples in the Chichimec Sea. Carbondale: Southern Illinois University Press.

Palkovich AM (1985) Historic population of the eastern Pueblos: 1540–1910. Journal of Anthropological Research *41*:401–426.

Reff DT (1987) The introduction of smallpox in the greater Southwest. American Anthropologist *89*:704–708.

Reff DT (1991) Disease, Depopulation, and Culture Change in Northwest New Spain, 1518–1764. Provo: University of Utah Press.

Reff DT (1992) Contact shock in Northwestern New Spain, 1518–1764. In JW Verano and DH Ubelaker (eds.): Disease and Demography in the Americas. Washington, DC: Smithsonian Institution Press, pp. 265–276.

Riley C (1986) An overview of the greater Southwest in the protohistoric period. In F Mathien and R McGuire (eds.): Ripples in the Chichimec Sea. Carbondale: Southern Illinois University Press, pp. 45–54.

Riley C (1987) The Frontier People: The Protohistoric Period in the Greater Southwest. Albuquerque: University of New Mexico Press.

Schroeder AH (1979) Pueblos abandoned in historic times. In A Ortiz (ed.): Handbook of North American Indians, vol. 9, Southwest. Washington, DC: Smithsonian Institution Press, pp. 236–254.

Simmons M (1966) New Mexico's smallpox epidemic of 1780–1781. New Mexico Historical Review *41*:319–324.

Simmons M (1979) History of Pueblo-Spanish relations to 1821. In A Ortiz (ed.): Handbook of North American Indians, vol. 9, Southwest. Washington, DC: Smithsonian Institution Press, pp. 178–193.

Spielmann KA (1983) Late prehistoric exchange between the Southwest and the Southern Plains. Plains Anthropologist *28*:257–272.

Spielmann KA (1989) Colonists, hunters and farmers: Plains Pueblo interaction in the seventeenth century. In DH Thomas (ed).: Columbian Consequences, vol. 1. Washington, DC: Smithsonian Institution Press, pp. 101–114.

Starr F (1899) A study of a census of the Pueblo of Cochiti, New Mexico. Proceedings of the Davenport Academy of Natural Sciences *7*: 533–544.

Upham S (1982) Polities and Power. New York: Academic Press.

Upham S (1984) Adaptive diversity and Southwestern abandonment. Journal of Anthropological Research *40*:235–256.

Upham S (1986) Smallpox and climate in the American Southwest. American Anthropologist *88*: 115–128.

Upham S (1987) Understanding the disease history of the Southwest: A reply to Reff. American Anthropologist *89*:708–710.

Upham S (1992) Population and Spanish contact in the Southwest. In JW Verano and DH Ubelaker (eds.): Disease and Demography in the Americans. Washington, DC: Smithsonian Institution Press, pp. 223–236.

Chapter 9

Bioarchaeological Investigations of Protohistoric Pueblo Health and Demography

Ann L.W. Stodder

Paul H. Rosendahl, Ph.D., Inc., Cultural Resources Management Studies and Services, G.M.F., Guam 96921

The bioarchaeological analysis of skeletons from two large villages occupied prior to and during the first decades of Spanish colonization in New Mexico forms the basis of this chapter. One of these settlements is San Cristobal, a Tano site in the Galisteo Basin at the eastern fringe of Pueblo territory, and the other is Hawikku, a Zuni village in western New Mexico. Fray Marcos de Niza's journey to the legendary golden city of Cibola (Hawikku) in 1539 began an era of profound and as yet incompletely understood political, religious, and demographic upheaval for the Pueblos (see also Palkovich, this volume). Bioarchaeological data from the protohistoric and prehistoric periods provide otherwise unobtainable information on the demography and health status of Pueblo populations at this critical time in their history. Such data also serve as a basis for evaluating the biological impact of European contact and colonization on these people.

Despite the arrival of the Spanish, the osteological picture is one of long-term continuity in some aspects of Pueblo health, such as the frequencies of anemia and dental pathology. However, contact period samples have notably high frequencies of skeletal lesions from infections and traumatic injuries. Differences in the nature of skeletal infection and in the patterns of subadult morbidity in the San Cristobal and Hawikku skeletal samples also suggest that there were health differences in the Eastern and Western Pueblo populations in this period.

EARLY CONTACT

The early contact period, 1539–1598, was one of intermittent contact by Franciscan missionaries and would-be conquerors. Food was demanded and obtained by force when necessary by these intruders. Coronado and his army unsuccessfully laid seige to Hawikku for several months in 1540, massacred the inhabitants of Tiguex, and destroyed about a dozen other villages. He was later indicted on six counts of excessive brutality for this behavior (Hammond and Rey, 1940). Although not all of the later encounters were confrontational, their chronicles share themes of their economically extractive policies.

Little written documentation remains from early colonial New Mexico because al-

In the Wake of Contact: Biological Responses to Conquest, pages 97–107

most 100 years of ecclesiastical and civilian government records were destroyed in the Pueblo Revolt of 1680. This pan-Pueblo uprising succeeded in driving the Spaniards out of the Pueblo region until the final battles of reconquest in 1696.

By 1640, some 25 Franciscan churches had been built by converted Indians (Forrestal, 1954). Native religious practices were forbidden, kivas were burned, and political and religious factionalism between and within Pueblos heightened (Snow, 1981; Wilcox, 1991). Spanish interference with trade between Plains hunters and Pueblo farmers exacerbated inter-Pueblo feuding in the east (Spielmann, 1991).

Colonists established ranches in prime agricultural lands granted under the Spanish *encomienda* system (Simmons, 1969). Tribute in the form of labor, maize, buffalo hides, cotton *mantas*, and other manufactured items were collected from every Pueblo household (Scholes, 1935). Demand for goods combined with the drain on labor and land disrupted household, local, and regional economies and subsistence practices (Spielmann, 1989). Moreover, the late 1500s and 1600s were a time of drought, and natives and Europeans alike suffered from famine (Scholes, 1935; Hackett, 1937).

Population and Disease

The first recorded epidemic, thought to be either smallpox (Cook, 1946; Forrestal, 1954) or scarlet fever (Dobyns, 1988), occurred in the 1630s. This epidemic may not have been the first to reach the Pueblos. Dobyns (1983) proposes that a hemispheric pandemic of smallpox spread to the Pueblos in 1520–1524, and Upham (1986) suggests that smallpox had arrived in the late 1500s, prior to colonization. Jesuit missionaries traveling through northwest Mexico in the late 1500s observed mass graves in native villages, and rituals devoted to deities representing disease symptoms—indirect evidence of the spread of epidemic disease through native populations in advance of direct European contact (Reff, 1985). This type of evidence remains to be demonstrated for the Pueblo area. Franciscan records after the Pueblo reconquest document multiple epidemics beginning with smallpox at Nambe in 1719 (Chavez, 1957; Stodder, 1990). Diseases depopulated the area from that time through the eighteenth century. Pueblo populations reached their nadir of 12,000 to 13,000 in about 1750 (Simmons, 1979).

Demographic disruptions also took the form of multiple migrations and settlement shifts at local and regional levels throughout the 1600s. By 1640, Schroeder (1972) estimates that as many as half of the Rio Grande Pueblos had been abandoned. Many more settlements, San Cristobal and Hawikku among them, were abandoned at the time of the revolt. Forcible Spanish resettlement after 1700 futher reduced the number of settlements and consequently increased distances between them (Wilcox, 1981).

Ubelaker (1988:292) has estimated precontact (ca. 1500) Southwestern Indian population size at 454,200, with the most extreme decline, 35%, occurring between 1600 and 1700. Upham (1992) suggests at least a 50% population decline in the Rio Grande Pueblos between 1500 and 1600, from 132,000 to 70,000 people. Additional, fine-grained studies of intrasite and site cluster chronologies are needed to understand the growth and decline of specific communities in relation to local and regional population trends both before contact and prior to the revolt.

Large multicomponent sites such as San Cristobal and Hawikku are typical of the protohistoric period (Creamer, 1991) and illustrate the problem of deriving regional population estimates from sites with rapidly changing demographic histories. San Cristobal was one of nine or 10 Galisteo Pueblos in the late 1400s, but several were abandoned after Teya raids in the 1520s (Nelson, 1914). This abandonment was part of long-term decline from prominence as a center of glazeware manufacture and trade. Wilcox (1991:152) attributes their eclipse by other peoples to increased competition with nearby Pecos Pueblo for agricultural resources and Plains trade. San Cristobal appears to have reached its population peak prior to contact during this era of general population decline (Lang, 1977).

Hawikku appeared to maintain a large number of inhabitants throughout its occupation from about 1425 to 1680 (Kintigh, 1985), but a Zuni-wide population decline by as much as 54% by 1581 is indicated archaeologically (Kintigh, 1988). After a church at Hawikku was built in 1629, the village's increasing unrest led to abandonment and reoccupation, possibly twice, before the revolt (Hodge, 1937; Ferguson, 1981).

HUMAN SKELETAL REMAINS

Skeletal Samples

San Cristobal was excavated in the early part of this century by Nels Nelson (1914, 1916) and by Louis Sullivan, a young protégé of Aleš Hrdlička. Except for Nelson's two brief reports, no results of the work at San Cristobal were published. However, extensive field records are available, most importantly Sullivan's (n.d.) detailed list of the age and sex of all exhumed individuals. Of the 492 burials recorded in the field records, 277 were saved, most of them from Sullivan's midden excavations. Selective curation reduced the percentage of subadults from 50% in the field records to only 34% in the collection (Table 1).

The burials from San Cristobal probably span the entire occupation of the site, ca. 1350–1680. However, a conservative interpretation of Nelson's field notes and recent surface collections allows the assignment of most skeletons to either precontact or postcontact components (19% prehistoric and 48% historic) (Stodder, 1990). Although far from optimal, these assignments suggest that the San Cristobal skeletal collection is predominately comprised of late protohistoric and early contact skeletons. Because much of the contrast in the San Cristobal and Hawikku paleoepidemiological profiles is in the subadult fraction of the collections, it is of some consolation that of 83 dated subadults from San Cristobal, 77% are from historic contexts.

Hodge's records for Hawikku indicate that 317 cremations and 679 burials were found

TABLE 1. Age Distribution of Recorded Vs. Collected Burials

	San Cristobal				Hawikku			
	Field records		Analysed sample		Field records		Analyzed sample	
	N	%	N	%	N	%	N	%
Infants	95	19	25	9				
Children	132	27	65	23	363[a]	44	51	27
Adolescents	20	4	5	2	69	9	14	8
Adults	245	50	182	66	389	47	123	65
Total	492		277		821		188	

N = number of individuals; %, percentage of sample.
Infants, 0–11.9 months; children, 1–9.9 years; adolescents, 10–15.9 years.
[a]Hawikku infants and children are combined because they are not consistently distinguished in field records.

TABLE 2. Age Distribution of Analyzed Burials

	San Cristobal		Hawikku	
Age	N	%	N	%
<1	27	10	25	13
1–4.9	38	14	16	9
5–9.9	23	9	12	6
10–14.9	6	2	14	8
15–19.9	26	10	18	10
20–24.9	25	9	31	16
25–29.9	24	9	17	9
30–34.9	29	11	13	7
35–39.9	20	7	16	9
40–44.9	12	4	8	4
45–49.9	23	9	10	5
50+	8	3	8	4
Indeterminate[a]	7	3		
Total	268	100	188	100

N, number of individuals; %, percentage of sample.
[a]Adults of indeterminate age.

(Smith et al., 1966). Of these, 261 burials were collected. Like the San Cristobal collection, there is a substantial discrepancy in the representation of subadults versus adults in the recorded and collected assemblages. Fifty-three percent of the burials recorded at Hawikku were subadults, but subadults comprise only 35% of the saved and provenienced collection (Table 1). A detailed age distribution of the analyzed skeletal samples is shown in Table 2.

Of 188 analyzed Hawikku burials, 128 are datable based on the site's chronology. Of these burials, 67% came from the chapel or from mortuary features that yielded Matsaki wares, produced from 1475 to 1650 (Kintigh, 1985), later ceramics, or European artifacts. Although the dating of the entire assemblage is not optimal, the sample is predominantly protohistoric or early historic.

Anemia

Porotic hyperostosis and cribra orbitalia are ubiquitous in prehistoric Southwestern populations (Stodder and Martin, 1992), including the San Cristobal and Hawikku samples. Orbital, postorbital, or both types of lesions are present in 90% of 188 San Cristobal crania and 84% of 127 Hawikku crania. Prehistoric Anasazi populations with similarly high frequencies of cranial lesions include Black Mesa, 88% (Martin et al., 1991); Mesa Verde, 70% (Stodder, 1987); and Chaco Canyon, 72% (El-Najjar et al., 1976).

The frequency of active lesions in subadult crania is highest among 1–1.9 year olds (Table 3); thereafter, frequencies decline. Sample sizes for these age groups are small, but the frequency of unremodeled lesions in 10–15.9 year olds from Hawikku is significantly higher (Fisher's exact probability test, $p < 0.05$) than in adolescents from San Cristobal.

These lesions in the Anasazi are interpreted as resulting from iron-deficiency anemia caused by a combination of factors, including nutritional and infection status related to local ecology and community hygiene (Walker, 1985; Palkovich, 1987; Reinhard, 1988; Martin et al., 1991). Children with skeletal evidence of infections also show high porotic hyperostosis frequencies (92%, San Cristobal; 80%, Hawikku). Among the children who died, those with skeletal infections are extremely likely to have experienced anemia, but so are all others who died young.

Enamel Hypoplasia

The age patterns and frequencies of enamel defects in deciduous and permanent teeth exhibit consistent intersite differences suggesting heightened morbidity among Hawikku subadults. Hypoplastic enamel defects occur in the deciduous incisors of 30% (6/20) of the Hawikku children and in 3% (1/33) of the San Cristobal children. More individuals in the Hawikku sample (94% of 111) have one or more examples of hypoplasia in a permanent incisor or canine than San Cristobal individuals (85% of 146). Hawikku dentitions also have a sig-

TABLE 3. Frequency of Active Porotic Hyperostosis in Subadults

	San Cristobal			Hawikku		
Age	No. with lesions	N	%	No. with lesions	N	%
0–1	6	11	55	5	12	42
1–1.9	14	18	78	7	8	88
2–4.9	4	14	29	2	4	50
5–9.9	5	19	26	3	6	50
10–15.9	0	5	0	2	9	22*
Total	28	66	42	20	40	50

N, number observable; %, percentage with lesions.
*Fisher's exact probability test, $p \leqslant 0.05$.

nificantly greater average number of enamel hypoplasias per individual, 6.62, than their San Cristobal counterparts, 4.40 (χ^2, $p < 0.001$). The frequency of defects in each of the permanent incisors and canines is consistently higher in the Hawikku dentitions (Stodder, 1990).

The frequencies of enamel hypolasias in six-month age intervals for the maxillary central incisors and mandibular canines are shown in Figures 1 and 2. The number of individuals represented in the teeth studied are as follows: incisors—San Cristobal 107, Hawikku 72; canines—San Cristobal

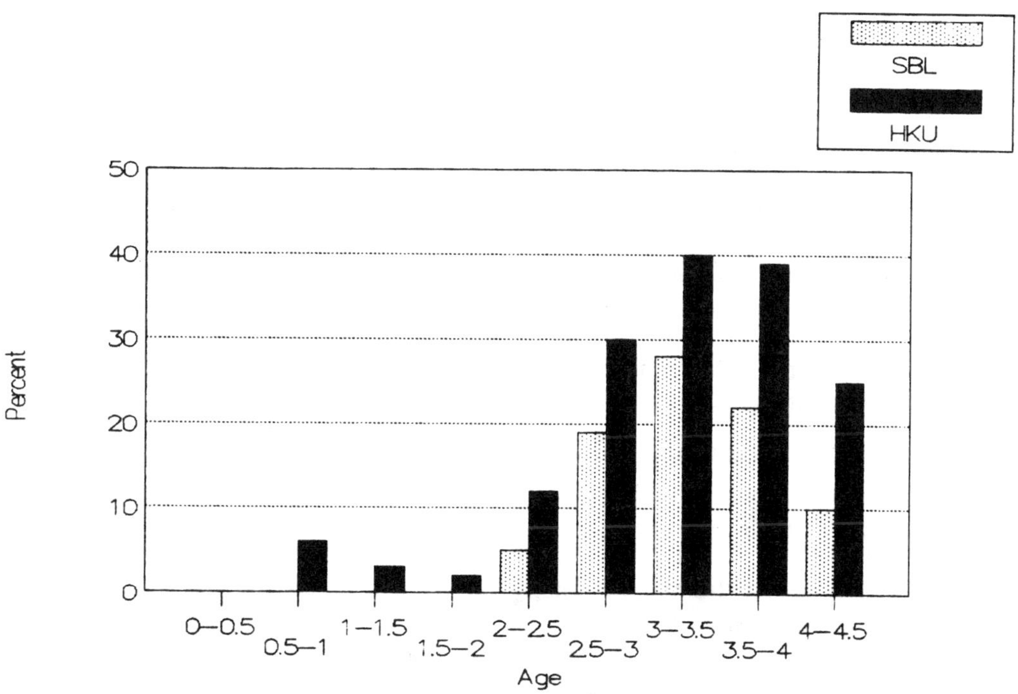

Fig. 1. Enamel hypoplasia formation, maxillary central incisors (HKU, Hawikku; SBL, San Cristobal).

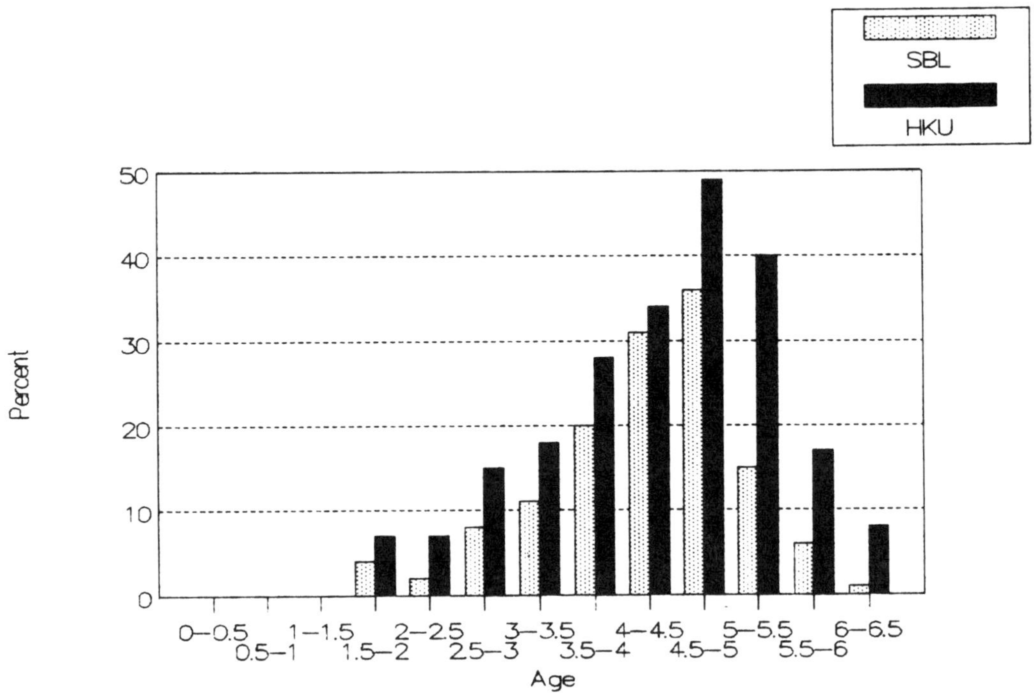

Fig. 2. Enamel hypoplasia formation, mandibular canines (HKU, Hawikku; SBL, San Cristobal).

103, Hawikku 86. Age calibration is based on averages for unworn Anasazi tooth crown heights (Stodder, 1990) and a standard formation chronology in Massler et al. (1941). Although revisions of the Massler chronology are needed (Skinner and Goodman, 1992), the overall patterns are important. The frequencies of defects are greater for every six-month age interval in both teeth for the Hawikku specimens than they are for the San Cristobal specimens.

Skeletal Infection

Evidence of skeletal infection is present in 23% (48/210) of San Cristobal and in 36% (51/142) of the Hawikku skeletons. These are substantially higher frequencies than those reported for most prehistoric Anasazi collections (Stodder and Martin, 1992), but are approached by Black Mesa (23%), where infections were characterized as predominately mild periostitis (Martin et al., 1991), and Chaco Canyon (17%) (Akins, 1986). Tuberculosis and a treponemal disease, endemic but infrequently observed in the Anasazi area (El-Najjar, 1979), are represented by lesions in the Hawikku and San Cristobal collections (Stodder, 1990).

Frequencies of skeletal infection in adults from the two sites are similar, as are the figures for San Cristobal adult and subadult skeletal involvement (Table 4). The high frequencies of skeletal infection in Hawikku subadults, infants especially, differentiates the two sites.

Most skeletal lesions attributed to infection in these samples were classified according to 1) their dominant morphology (proliferative or osteolytic [destructive]) and 2) their distribution (localized in a single body segment [for example, upper limb], or systemically distributed). Proliferative lesions observed in these samples include periostitis, osteitis, and suites of lesions attributed

TABLE 4. Age Distribution of Skeletal Infection

	San Cristobal			Hawikku		
Age	No. with lesions	N	%	No. with lesions	N	%
Subadults						
0–11.9 months	4	16	25	15	20	75**
1–1.9 years	2	16	12	3	8	37
2–4.9	4	11	36	3	6	50
5–9.9	4	19	21	6	11	54
10–15.9	1	5	20	2	12	17
0–15.9	15	67	22	29	57	51**
Adults						
16–25.9	11	49	22	7	32	22
26–35.9	12	45	27	6	24	25
36+	7	47	17	9	29	31
Indeterminate age	2	2	100	0	0	0
16+	32	143	23	22	85	26
Total	48	212	23	51	142	36*

N, number observable; %, percentage with lesions.
χ^2 significance levels: $*p < 0.05$; $**p < 0.001$.

to treponemal infection. Osteolytic lesions include nonspecific destructive lesions and those attributed to tuberculosis. Comparison of the frequencies of proliferative versus osteolytic lesions in the San Cristobal and Hawikku collections (Table 5) suggests a fundamental difference in the nature of skeletal infection in the communities.

Infracranial lesions attributed to treponematosis, including diaphyseal expansion and bowing, localized expansive nodes, coarse striations and pitting, are present in

TABLE 5. Classification of Skeletal Infections in the San Cristobal and Hawikku Skeletal Samples

	San Cristobal		Hawikku	
	% of cases	% of sample	% of cases	% of sample
Proliferative lesions	58	15	89**	34**
Localized	42	11	44	17
Systemic	11	3	40*	15*
Treponematosis	5	1	5	2
Osteolytic lesions	34**	9	9	4
Localized	20	5	5	2
Tuberculosis	14	4	4	1
Other infections	8	2	2	<1
Total	55	212	55	142

% of cases, the percentage of infections per category; % of sample, the percentage of individuals in sample with infections in each category.
χ^2 significance levels: $*p < 0.01$, $**p < 0.001$.

three adults and two subadults from each site. No stellate scars or circumvallate lesions are evident in the crania, but fewer than half of the endemic treponematosis cases reported from the Southwest show cranial involvement (Stodder, 1990).

Localized periostitis affects similar proportions of the two site samples, but systemic periostitis, some of which probably represents additional cases of treponemal infection, is considerably more abundant among the Hawikku burials. Systemic periostitis is present in 67% of 57 Hawikku subadults compared to 7% of 67 San Cristobal subadults. Of those who died in the first year of life at Hawikku, 55% ($n = 20$) are affected, as compared to 19% ($n = 16$) of San Cristobal infants.

In contrast, a significantly greater proportion of systemic infections in the San Cristobal burials are characterized by osteolytic lesions. These are diagnosed as tuberculosis in five adults and three children from San Cristobal and in one child and one adult from Hawikku (Stodder, 1990). In adults, tuberculous lesions are located in the ribs, thoracic and lumbar vertebrae, sacrum, innominate, hip, and foot. Subadults exhibit marked diaphyseal expansion and osteolytic foci at metaphyses in the upper and lower limbs as well as in ribs, vertebrae, and metacarpals (Stodder, 1990). The five tuberculous adults from San Cristobal were all interred in Midden E, one of seven extramural burial areas at the site. The child cases are from the adjacent Midden D. This clustering suggests the possibility of a substantial increase in tuberculosis during the later stages of occupation of San Cristobal. Tuberculosis is an opportunistic disease that tends to be associated with times of famine, warfare, and high population mobility (Youmans, 1979).

Cranial Trauma

There is a marked increase in the frequencies of reported cranial trauma throughout the late prehistoric period in the Rio Grande Pueblos (Table 6). At San Cristobal and Hawikku, males seemed to be particularly likely to experience trauma suggestive of interpersonal violence. At San Cristobal, 20% of 66 adult male crania show trauma, and only 11% of 65 female crania are affected. Corresponding figures for Hawikku are 17% of 40 and 7% of 53 crania in males and females, respectively.

CONCLUSION

Ethnohistoric and archaeological data indicate that in addition to introduced epidemic disease, other factors including warfare, population displacement, drought, famine, and large-scale disruption of the Pueblo economic system contributed to the precipitous decline in the sixteenth- and seventeenth-century Pueblo populations. Despite the limitations inherent in these skeletal series, the skeletal data from San Cristobal and Hawikku demonstrate an additional, and perhaps more subtle, process of depopulation—namely, an increase in indigenous infections in an era of demographic, political, and economic instability.

High frequencies of cranial trauma in adult males reflect the increase in armed confrontation with the Spaniards, as well as hostilities among Pueblos and between Pueblo and non-Pueblo natives in the protohistoric and the early contact periods (Spielmann, 1991; Wilcox, 1991). The ubiquity of porotic hyperostosis suggests a serious and widespread impact on nutritional status.

Some researchers (e.g., Wood et al., 1992) suggest that because enamel hypoplasia formation takes place following resolution of physiological stress, a greater frequency of these defects may indicate a healthier population. Nevertheless, the skeletal infections in 75% of the infants clearly indicate health problems for children at Hawikku.

An increase of endemic treponematosis is suggested by the frequency of systemic periostitis. However, in the absence of con-

TABLE 6. Reported Frequencies of Cranial Trauma in Adults from Late Prehistoric Skeletal Samples

Site	Dates	No. of adults	Cranial trauma (%)	References
Tijeras	1300–1425	64	2	Ferguson, 1980
Arroyo Hondo	1300–1450	47	0	Palkovich, 1980
Cochiti	1300–1550	122	2	Heglar, 1974
Paa'ko	1300–1600	57	3	Ferguson, 1980
Pecos	1425–1550	204	3	Hooton, 1930
Pecos	1550–1600	93	8	Hooton, 1930
Hawikku	1425–1680	93	12	Stodder, 1990
San Cristobal	1350–1680	131	15	Stodder, 1990
Pecos	1600–1800	88	16	Hooton, 1930

vincing cases of venereal treponemal infection (syphilis) in pre-eighteenth-century skeletons (Baker and Armelagos, 1988), its presence has less ominous implications than tuberculosis. Nonvenereal treponemal diseases cause suffering, impair work ability, and may produce disabling deformities and reduce fertility, but unlike tuberculosis they tend not to be fatal without superinfection (Perine et al., 1984).

Dunnell (1991) applies the concepts of biological and cultural founder effect to the quantitative reduction and resultant condensing of postcontact native cultural systems; cultural traits remain, but the systems are altered. The timing and dimensions of Pueblo cultural disruption and alteration at the hands of Spanish explorers and colonists cannot be fully understood until the scale and complexity of precontact Pueblo society is investigated. The validity and usefulness of the paleoepidemiological data and of other data sets must be enhanced by further attention to the archaeological details of protohistoric population dynamics and biocultural adaptation both before and after Spanish contact. This is essential if we are to go beyond the stage of modeling and hemispheric speculation in our approach to population collapse, and if we are to understand the cultural systems whose history we seek to trace.

ACKNOWLEDGMENTS

This research was supported by a Richard H. Lounsbury Fellowship for Predoctoral Research in Anthropology from the American Museum of Natural History, by Sigma Xi, and by the University of Colorado Graduate School Foundation.

REFERENCES CITED

Akins NJ (1986) A Biocultural Approach to Human Burials from Chaco Canyon, New Mexico. Santa Fe: Reports of the Chaco Center 9.

Baker BJ, and Armelagos GJ (1988) The origin and antiquity of syphilis: Paleopathological diagnosis and interpretation. Current Anthropology *29*:703–720.

Chavez Fr. A (1957) Archives of the Archdiocese of Santa Fe. Publications of the Academy of American Franciscan History, Bibliographic Series, 8.

Cook SF (1946) The incidence and significance of disease among the Aztecs and related tribes. The Hispanic American Historical Review *26*:320–335.

Creamer W (1991) Developing complexity in the American Southwest: Developing a model for the Rio Grande Valley. Paper presented, Society for American Archaeology, Pittsburgh.

Dobyns HF (1983) Their Number Become Thinned. Knoxville: University of Tennessee Press.

Dobyns HF (1988) Discussant. The Southwest Symposium. Arizona State University, Tempe.

Dunnell RC (1991) Methodological impacts of catastrophic depopulation on American archaeology and ethnology. In DH Thomas (ed.): Columbian

Consequences, vol. 3, The Spanish Borderlands in Pan-American Perspective. Washington, DC: Smithsonian Institution Press, pp. 561–580.

El-Najjar MY (1979) Human treponematosis and tuberculosis: Evidence from the New World. American Journal of Physical Anthropology *51*: 599–618.

El-Najjar MY, Ryan D, Turner C, and Lozoff B (1976) The etiology of porotic hyperostosis among the prehistoric and historic Anasazi Indians of the southwestern United States. American Journal of Physical Anthropology *44*:477–488.

Ferguson C (1980) Analysis of human skeletal remains. In LS Cordell (ed.): Tijeras Canyon, Analyses of the Past. Albuquerque: University of New Mexico Press, pp. 121–148.

Ferguson TJ (1981) The emergence of modern Zuni culture and society: A summary of Zuni tribal history, A.D. 1450–1700. In DR Wilcox and WB Masse (eds.): The Protohistoric Period in the American Southwest, A.D. 1450–1700. Arizona State University Anthropological Research Papers 24, pp. 336–353.

Forrestal P (1954) Benavides' Memorial of 1630. Washington, DC: Academy of American Franciscan History.

Hackett CW (1937) Historical Documents Relating to New Mexico, Nueva Vizcaya, and Approaches Thereto, to 1773, vol. 3. Washington, DC: Carnegie Institution.

Hammond G, and Rey A (1940) Narratives of the Coronado Expedition, 1540–1542. Coronado Historical Series, vol. 1. Albuquerque: University of New Mexico Press.

Heglar R (1974) The Prehistoric Population of Cochiti and Selected Interpopulational Comparisons. Ph.D. dissertation, Department of Anthropology, University of Michigan, Ann Arbor.

Hodge FW (1937) History of Hawikuh. Los Angeles: The Southwest Museum.

Hooton EA (1930) The Indians of Pecos Pueblo. New Haven: Yale University Press.

Kintigh K (1985) Settlement, Subsistence, and Society in Late Zuni prehistory. Anthropological Papers of the University of Arizona 44.

Kintigh K (1988) Protohistoric Transitions in the Western Pueblo Area. Paper presented, Southwest Symposium, Arizona State University, Tempe.

Lang R (1977) Archaeological Survey of the Upper San Cristobal Arroyo Drainage, Galisteo Basin, New Mexico. School of American Research Contract Archaeology Program Report 37.

Martin DL, Goodman AH, Armelagos GJ, and Magennis AL (1991) Black Mesa Anasazi Health: Reconstructing Life from Patterns of Death and Disease. Southern Illinois University at Carbondale, Center for Archaeological Investigations, Occasional Paper 14.

Massler M, Schour I, and Poncher HG (1941) Developmental pattern of the child as reflected in the calcification pattern of the teeth. American Journal of Diseases of Children *62*:33–67.

Nelson NC (1914) Pueblo ruins of the Galisteo Basin, New Mexico. Anthropological Papers of the American Museum of Natural History 15 (part 1).

Nelson NC (1916) Chronology of the Tano ruins, New Mexico. American Anthropologist *18*:159–180.

Palkovich AM (1980) Pueblo Population and Society: The Arroyo Hondo Skeletal and Mortuary Remains. Santa Fe: School of American Research Press.

Palkovich AM (1987) Endemic disease patterns in paleopathology: Porotic hyperostosis. American Journal of Physical Anthropology *74*:527–538.

Perine PL, Hopkins DR, Niemel PLA, St. John RK, Causse G, and Antal GM (1984) Handbook of Endemic Treponematoses: Yaws, Endemic Syphilis, and Pinta. Geneva: World Health Organization.

Reff DT (1985) The Demographic and Cultural Consequences of Old World Disease in the Greater Southwest, 1520–1660. Ph.D. dissertation, Department of Anthropology, University of Oklahoma, Norman.

Reinhard KJ (1988) Cultural ecology of prehistoric parasitism on the Colorado Plateau as evidenced by coprology. American Journal of Physical Anthropology *77*:355–366.

Scholes FV (1935) Civil government and society in New Mexico in the seventeenth century. New Mexico Historical Review *10*:71–111.

Schroeder A (1972) Rio Grande ethnohistory. In A Ortiz (ed.): New Perspectives on the Pueblos. Albuquerque: University of New Mexico Press, pp. 42–70.

Simmons M (1969) Settlement patterns and village plans in colonial New Mexico. Journal of the West *8*:7–21.

Simmons M (1979) History of Pueblo-Spanish relations to 1821. In A Ortiz (ed.): Handbook of North American Indians, vol. 9, Southwest. Washington, DC: Smithsonian Institution Press, pp. 178–193.

Skinner M, and Goodman AH (1992) Anthropological uses of developmental defects of enamel. In S Saunders and MA Katzenberg (eds.): Skeletal Biology of Past Peoples: Research Methods. New York: Wiley-Liss, pp. 153–174.

Smith W, Woodbury R, and Woodbury N (1966) The excavation of Hawikuh by Frederick Webb Hodge. Report of the Hendricks-Hodge Expedition. Contributions of the Museum of the American Indian, Heye Foundation, 20.

Snow DH (1981) Protohistoric Rio Grande economics: A review of trends. In DR Wilcox and WB Masse (eds.): The Protohistoric Period in the American Southwest, AD 1450–1700. Arizona State University Anthropological Research Papers 24, pp. 354–377.

Spielmann KA (1989) Colonists, hunters, and farmers: Plains-Pueblo interaction in the seventeenth century. In DH Thomas (ed): Columbian Consequences, vol. 1, Archaeological and Historical Perpsectives on the Spanish Borderlands West. Washington, DC: Smithsonian Institution Press, pp. 101–114.

Spielmann KA (1991) Coercion or cooperation? Plains-Pueblo interaction in the protohistoric period. In KA Spielmann (ed): Farmers, Hunters, and Colonists: Interaction Between the Southwest and the Southern Plains. Tucson: University of Arizona Press, pp. 36–50.

Stodder ALW (1987) The physical anthropology and mortuary behavior of the Dolores region Anasazi. In KL Petersen and JD Orcutt (eds.): Dolores Archaeological Program Supporting Studies: Settlement and Environment. Denver: U.S. Bureau of Reclamation Engineering and Research Center, pp. 309–504.

Stodder ALW (1990) Paleoepidemiology of Eastern and Western Pueblo Communities in Protohistoric New Mexico. Ph.D. dissertation, Department of Anthropology, University of Colorado, Boulder.

Stodder ALW, and Martin DL (1992) Health and disease in the Southwest before and after Spanish contact. In JW Verano and DH Ubelaker (eds.): Disease and Demography in the Americas. Washington, DC: Smithsonian Institution Press, pp. 55–73.

Sullivan, L (n.d.) Notes on the excavation of San Cristobal Pueblo. On file, Department of Anthropology, American Museum of Natural History, New York.

Ubelaker DH (1988) North American Indian population size, A.D. 1500 to 1985. American Journal of Physical Anthropology 77:289–294.

Upham S (1986) Smallpox and climate in the American Southwest. American Anthropologist 88: 115–128.

Upham S (1992) Population and Spanish contact in the Southwest. In JW Verano and DH Ubelaker (eds.): Disease and Demography in the Americas. Washington, DC: Smithsonian Institution Press, pp. 223–236.

Walker PL (1985) Anemia among the prehistoric Indians of the American Southwest. In CF Merbs and RJ Miller (eds.): Health and Disease in the Prehistoric Southwest. Arizona State University Anthropological Research Papers 34, pp. 139–162.

Wilcox DR (1981) Changing perspectives on the protohistoric Pueblos, A.D. 1450–1700. In DR Wilcox and WB Masse (eds.): The Protohistoric Period in the American Southwest, A.D. 1450–1700. Arizona State University Anthropological Research Papers 24, pp. 378–409.

Wilcox DR (1991) Changing contexts of Pueblo adaptations, A.D. 1250–1600. In KA Spielmann (ed.): Farmers, Hunters, and Colonists: Interaction Between the Southwest and the Southern Plains. Tucson: University of Arizona Press, pp. 128–154.

Wood JW, Milner GR, Harpending HC, and Weiss KM (1992) The osteological paradox. Current Anthropology 33:343–370.

Youmans GP (1979) Tuberculosis. Philadelphia: WB Saunders.

Chapter 10

The Decline of the Chumash Indian Population

Phillip L. Walker and John R. Johnson

Department of Anthropology, University of California, Santa Barbara, CA 93106 (P.L.W.); Department of Anthropology, Santa Barbara Museum of Natural History, Santa Barbara, CA 93105 (J.R.J.)

In many respects, the catastrophic consequences that European contact had for the Chumash Indians of southern California are similar to those experienced by other Native American groups. Infectious diseases were introduced and rapidly spread to the local Indians. Epidemics were accompanied by massive social disruption and eventually resulted in decimation of the native population.

Often this process occurred in native villages beyond the purview of European colonists. For many North American tribes, a few passages in explorers' diaries are all that documents their existence. The demographic and social collapse of the Chumash, in contrast, is exceptionally well documented by records in mission archives. These historical documents provide an unusually vivid picture of the demographic processes responsible for the decline in Native American populations following the arrival of Europeans in the New World.

Although the Chumash were first visited by Spanish explorers in 1542, they had only a few encounters with Europeans during the next 200 years (Table 1). Beginning in 1772, Spanish soldiers and missionaries started to build presidios (garrisoned forts) and missions along the Santa Barbara Channel in the heart of Chumash territory (Fig. 1). In their dealings with the Chumash, the Spanish used their *reducción* strategy of removing Indians from native villages and concentrating them at missions. As a result, nearly all of the Chumash who survived early epidemics moved to the missions. At the missions, they received religious and occupational instruction and served as labor for mission construction and agricultural activities (Hornbeck, 1983:46; Hoover, 1989:398).

MISSION RECORDS

The priests at the five missions in Chumash territory—San Luis Obispo, San Buenaventura, Santa Barbara, La Purísima, and Santa Inés—kept several kinds of records on the Indians they controlled. When an Indian entered the mission system, his or her name, age, and village of origin were recorded in the mission baptismal register. Often additional personal details were recorded, such as parent names, physical infirmities, and so on. Priests also maintained registers of marriages and deaths at each mission. They supplemented these records

In the Wake of Contact: Biological Responses to Conquest, pages 109–120

TABLE 1. Chronology of Important Epidemics and Other Events Associated with the European Colonization of the Chumash Territory

Year	Event
1542	Juan Rodríguez Cabrillo first contacts the Chumash
1587	Pedro de Unamuno lands near Morro Bay
1595	Sebastián Rodríguez Cermeño visits San Luis Obispo Bay
1602	Sebastián Vizcaíno visits the Santa Barbara Channel
1772	Mission San Luis Obispo founded
1782	Mission San Buenaventura and Santa Barbara Presidio founded
1769–1770	Portolá passes through Chumash territory
1776	Anza Expedition
1786	Mission Santa Bárbara founded
1787	Mission La Purísima Concepción founded
1796	Typhoid and pneumonia reported at missions
1801	Diphtheria epidemic
1804	Mission Santa Inés founded
1806	Measles epidemic
1821	Mexican independence from Spain
1822–1823	Measles epidemic
1824	Chumash uprising
1827–1828	Measles epidemic
1832	Unidentified epidemic, possibly influenza
1834–1836	Chumash missions secularized
1844	Smallpox epidemic spread from Monterey to La Purísima
1846	California is conquered by the United States

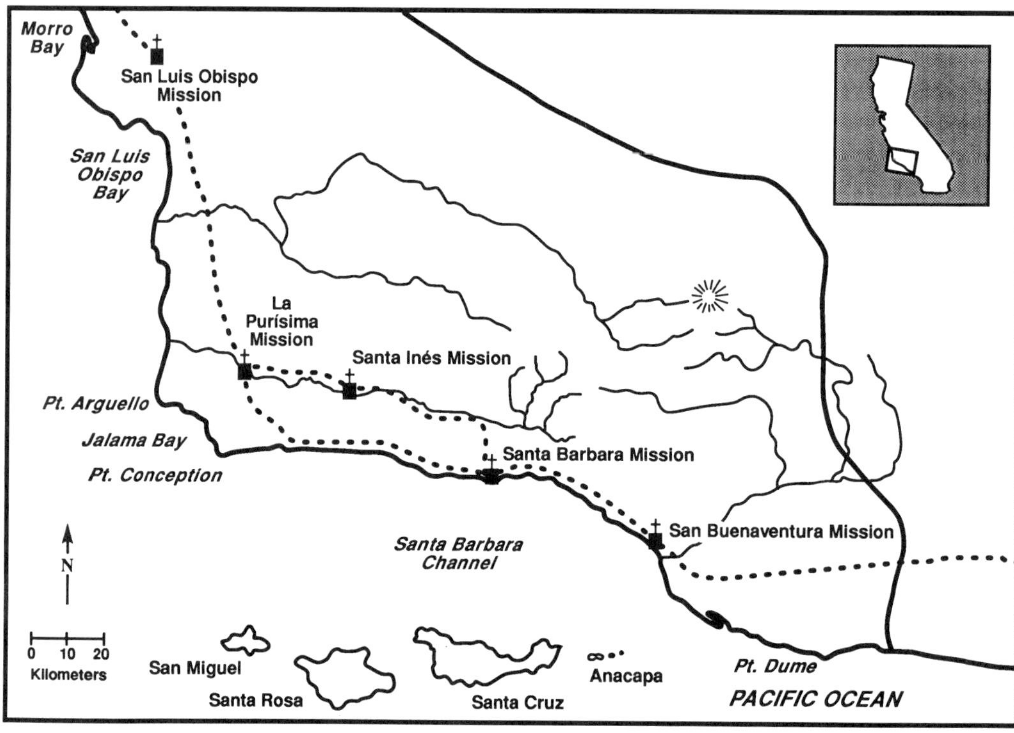

Fig. 1. Map of Spanish missions in the Chumash territory. The solid line marks the boundary of the area occupied by people speaking Chumash languages at the time of contact. The dotted line shows the principal routes of travel through Chumash territory during the mission period.

with *padrón* that contained information similar to the baptismal records and served as current censuses of the entire mission population (Johnson, 1988:48–50). Finally, the priest in charge of each mission was responsible for making annual reports that contained demographic data and agricultural statistics, including crop yields and livestock holdings. The annual reports provide a convenient source of information on the mission population size and the deaths that occurred each year. The reported statistics also can be used to explore the relationship between mission agricultural productivity and demographic variables such as population growth and death rate.

To learn more about the demographic processes responsible for the Chumash population decline, we cross-referenced baptismal and burial register entries (Johnson, 1988). This was done for each person baptized at Mission Santa Barbara between 1786 and 1828, at La Purísima between 1788 and 1851, and at Santa Inés between 1804 and 1844. Burial records were located for 8,476 of the Indians baptized at these missions; this total is about 90% of all baptisms recorded. Most missing burial entries are probably a result of deaths in native villages that escaped the missionaries' notice and deaths that occurred after the collapse of the mission system when systematic record keeping had ceased. A few additional people are probably unaccounted for because they died while fugitives from the mission system (Johnson, 1989:371).

POPULATION DECLINE

At the time of European contact, it is estimated that more than 15,000 Chumash were living in villages on the Northern Channel Islands and adjacent mainland (Brown, 1967; Johnson, 1982, 1988). It took about 50 years to relocate these people to missions. Relocation began in 1772 with the establishment of Mission San Luis Obispo, and it ended in 1822 when the last remaining Chumash villages were abandoned. Most people living on the densely populated mainland coast entered the mission system between 1787 and 1806. During the second decade of the nineteenth century, the priests turned their attention to converting the Chumash who remained in remote mainland interior and Channel Island villages.

The mission population reached its zenith in 1805 when 5,602 Indians were living at the five Chumash missions. During the next 30 years, deaths exceeded births and the Chumash population steadily declined. The only exception to this downward trend occurred between 1814 and 1816 when the mission population was temporarily augmented by an influx of people from the Channel Islands. By 1832, the last year for which accurate statistics are available, the mission Chumash population had been reduced to 1,182 people. In 1834, the Mexican government secularized the missions, and the living conditions of the survivors rapidly deteriorated. By the time of the first California state census in 1852, the Chumash had been reduced to fewer than 600 people (Johnson, 1993).

POPULATION STRUCTURE

A comparison of the Chumash population pyramids of 1782 and 1822 provides clues to the processes responsible for their decline (Fig. 2). In 1782, all of the Chumash in our sample lived in native villages. By 1822, all of these people had entered the mission system. There is some uncertainty in our reconstruction of the 1782 Chumash population because the baptismal and death registers omit people living in 1782 who died before baptism without ever entering the missions. Nevertheless, it provides a useful baseline for judging the demographic effects of missionization.

The 1782 population pyramid has two conspicuous features. First, there are more women than men. This biased adult sex ratio is especially apparent in the older age

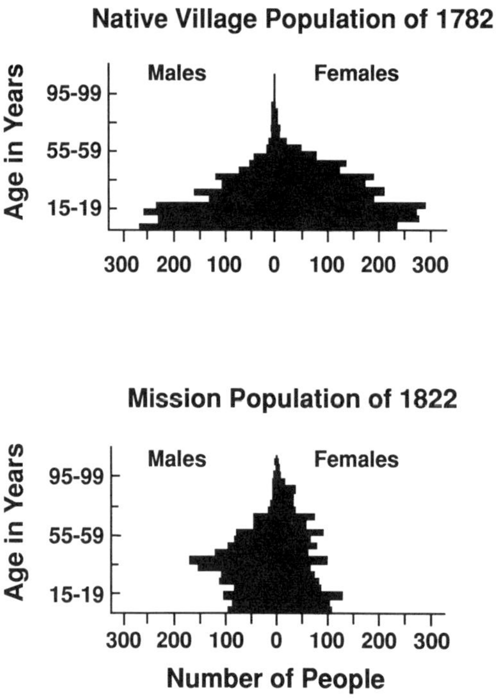

Fig. 2. Reconstructed population pyramids for the Chumash living in native villages in 1782 and the mission Chumash population in 1822.

groups. The paucity of older men in the 1782 population is possibly a result of warfare. Historical sources suggest that intervillage conflict was common before the founding of the missions (Engelhardt, 1932b:7; Brown, 1967:75–76; Burrus, 1967: 135; Geiger and Meighan, 1976:93, 113, 139; Johnson, 1988; Walker et al., 1989; Walker and Johnson, 1992). This appears to have been a continuation of a pattern of violence dating from before the arrival of Europeans (Walker, 1989; Lambert and Walker, 1991). Alternatively, it is possible that men were more reluctant than women to move to missions. As a result, more males than females would die in their villages without baptism. The latter interpretation, however, is inconsistent with the baptismal registers that show that early in the mission period, more men than women moved to the missions (cf. Coombs 1975).

The second striking characteristic of the pre-mission population pyramid is its somewhat constricted base (Fig. 2). Assuming this pattern is not a result of the priests failing to baptize the young, it suggests that in 1782 the Chumash had begun to experience either a low birth rate or high infant mortality. The latter seems more probable, especially considering the high birth rate of the Chumash who moved to the missions. It is possible that diseases introduced by Spanish explorers resulted in high infant mortality rates in the pre-mission Chumash population. Elsewhere in California at this time many Indian children were dying in epidemics (Palóu, 1913:213; Bolton, 1926:4, 161; Cook, 1976:18; Milliken, 1987:26).

The age structure of the mission Chumash in 1822 bears little resemblance to that of the villagers of 1782 (Fig. 2). The compressed base of the 1822 pyramid suggests that the reproductive problems of the pre-mission population were compounded during the mission period. There was a major change in the adult sex ratio. The 1822 population contains many more males than females. This is a complete reversal of the situation in 1782.

MISSION PERIOD EPIDEMICS

Mission death records show that six or seven times during the mission period death rates rose steeply for a few years and then decreased (Fig. 3; Walker and Johnson, 1992). During these mortality crises, as much as 15% of a mission's population died in a single year.

Often historical data are insufficient to identify with certainty the disease, or diseases, responsible for these periods of high mortality. In a few instances, however, it is possible to determine the cause of excessive mortality with a fair degree of certainty (Table 1). Between 1800 and 1802 an epidemic of what was probably diphtheria killed many Indians throughout California (Cook, 1976:19). The epidemic was first re-

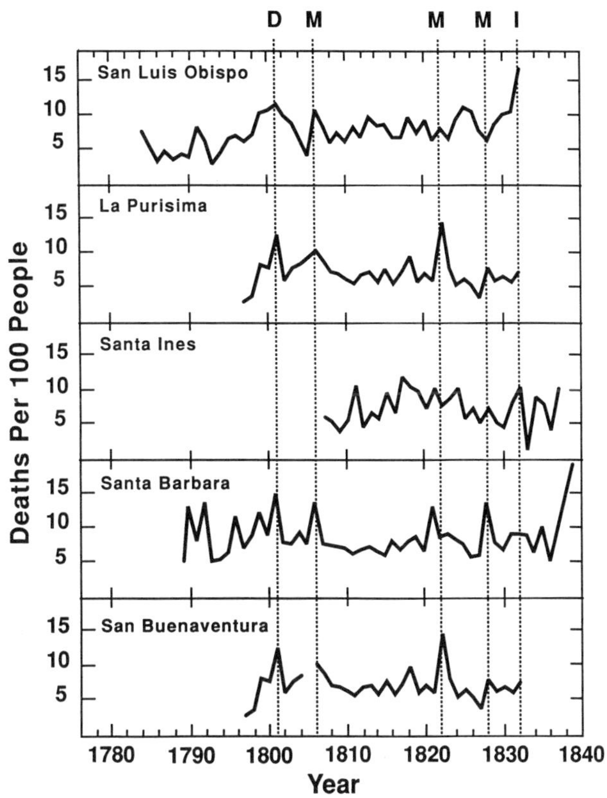

Fig. 3. Death rates at the Chumash missions. The letters above the vertical lines designate historically documented epidemics. D, diphtheria; M, measles; I, probably influenza.

ported in 1800 at Mission San Gabriel (Kenneally, 1965:2:176, 178, 264). By 1801, it had spread to the Chumash and mortality rates rose sharply at all of the missions (Fig. 3).

Measles first struck in 1806. This epidemic was particularly devastating because it occurred just after a major recruitment drive initiated in 1803. This conversion effort led most mainland Chumash to move to the missions (Engelhardt, 1930a:607; Johnson, 1989:369). Dense aggregations of Indians at the missions provided an ideal environment for the spread of measles. Many people contracted the disease and died. For example, at Mission Santa Barbara, 44 people died during a 15-day period (Cook, 1976: 19), and at Mission La Purísima more than 80 died (Cook, 1976:19; King, 1984). Measles was also responsible for later epidemics in 1821–1822 and 1827–1828 that claimed many lives (Engelhardt, 1930b: 60, 66).

Smallpox is not known to have hit the Chumash until 1844. Although it killed many of those who remained at San Luis Obispo and La Purísima, the epidemic did not reach Santa Inés (Storke, 1891:23; James, 1905:206; Engelhardt, 1932a:87; Walker and Hudson, 1993). The Chumash were not the only ones to suffer; people throughout much of California died from this widespread, but unevenly distributed, epidemic (Cook, 1939).

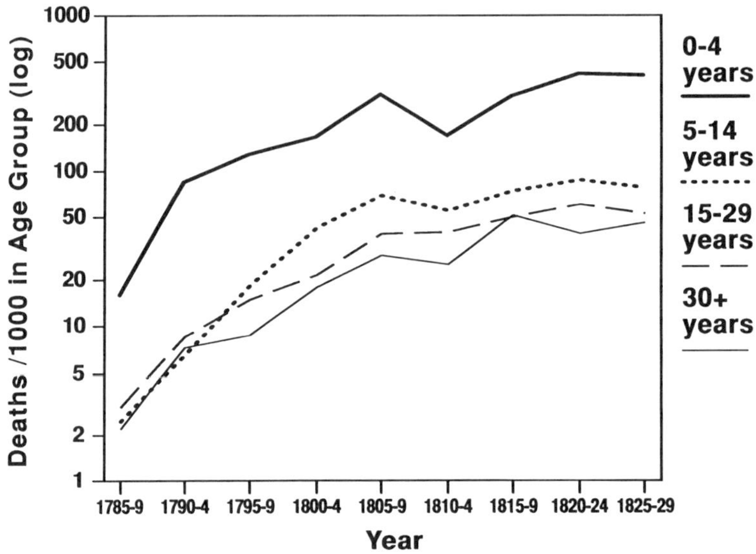

Fig. 4. Age-specific death rates plotted on a logarithmic scale. The values plotted 10 year averages for combined populations of Mission Santa Barbara, Santa Inés, and La Purísima.

Besides these short duration, high-mortality epidemics, the Chumash experienced the debilitating effects of chronic diseases such as syphilis and tuberculosis. Syphilis, which was rampant, reduced the reproductive capacity of the mission population and also their resistance to acute infections. The priests at San Luis Obispo wrote in 1814 that venereal disease was

> an infirmity with which generally all the Indians are infected to such an extent that any other illness during the various stages of the year kills them. Hence it is to my grief and to all who behold their unhappy fate that there are more deaths than births. (cited in Geiger and Meighan, 1976:75)

DEATH RATES

Although they fluctuated from year to year, there is a long-term trend toward increasing death rates (Figs. 2 and 4). During the first half of the mission period, 5.6% of the population died each year; during the last half, 7.7% died annually. Living conditions clearly became less favorable through time.

Increased interaction between the missions and the outside world may explain this trend toward higher death rates. As contacts with outsiders increased, so did the chances for transmission of fatal infectious diseases. During the first part of their existence, the missions were extremely isolated. This isolation resulted in part from prohibitions on foreign traders imposed by the Spanish government. Between 1790 and 1805, only 19 foreign ships are known to have visited the California coast (Ogden, 1975:156–159). Contact with traders increased after 1810. Spain's control over its holding in the New World was beginning to diminish, and a lucrative trade in sea otter pelts increased. Records show that between 1810 and 1821, 100 ships visited the California coast (Costello, 1990:69).

Considering their close spatial proximity and the regular travel of people between the missions along the Camino Real, there is a surprising lack of correspondence among the death rates of the five Chumash missions. Although a few major epidemics occurred almost simultaneously at all missions, the number of deaths would frequently rise sharply in some but not all Indian communities (Fig. 3). This discrepancy might reflect differences in the disease histories of each mission's population. For example, the Chumash at San Buenaventura and Santa Barbara were mainly from coastal villages that interacted intensively with the Spanish throughout the mission period. These people may have been exposed to and acquired immunities to diseases earlier than the inhabitants of La Purísima and Santa Inés, many of whom were recruited from small, remote inland villages (Johnson 1988:88–90).

There is no clear relationship between death rates and the economic success of the missions as reported in the priests' annual reports. We analyzed these records and found that Indians living at missions with low grain and livestock production suffered from mortality rates that were just as high as those at more productive missions. This pattern suggests that the causes of variation in mission death rates were unrelated to agricultural productivity. Perhaps this is because living conditions at a mission depended more on the priest in charge than they did on the economic achievements of the mission's Indian labor force. Some priests were excellent administrators and made good use of limited resources. Others, such as Padre Zalvidea at San Gabriel Mission, denied neophytes material comforts because they believed that "finery led Indians to run away" (Reid, 1978:273–274). Under his stewardship, the mission economy thrived while the Indians he controlled suffered miserably (Costello, 1992:79).

INFANT MORTALITY

High infant mortality was a significant factor contributing to the decline of the mission population. During the mission period the annual death rate for children less than five years old averaged 227 per 1,000 (Fig. 4). If a child managed to survive the first few years of life, the chance of living to adulthood increased markedly. The average annual mortality rate for children between 5 and 14 years of age was 26 per 1,000 or about one-tenth that of infants. For adults the average death rate was 34 per 1,000 for the 15–29 year old age group and 50 per 1,000 for people aged 30 years or older.

Although the exact causes of this extraordinarily high infant mortality are unclear, a lack of immunity to epidemic diseases was clearly important. For example, in 1806 nearly one-third of the children less than five years of age died during the measles epidemic (Fig. 5). Deaths from acute infections, however, provide only a partial explanation because infant mortality was nearly as high during normal years (Fig. 4). Other variables that presumably contributed to many infant deaths include crowded, unsanitary living conditions leading to the contamination of food and water, as well as the debilitating effects of congenital syphilis (Walker et al., 1989).

It is possible that infanticide also played a role. In 1794, Father Lasuen observed that the failure of the native population to grow was due to their

> . . . dominant vice of incontinence, and especially in the entire Channel region from Santa Barbara to San Luis, their inhuman and widespread practice of voluntary abortion, or on the part of the mother, of suffocating their newly born children. (Kenneally, 1965:378)

Although the priests may have exaggerated the frequency of infanticide because they considered it especially abhorrent, the existence of this practice is confirmed by elderly Chumash interviewed early in this century (Walker and Hudson, 1993: 107–108).

BIRTH RATE

The Chumash birth rate decreased significantly during the mission period (Fig. 6).

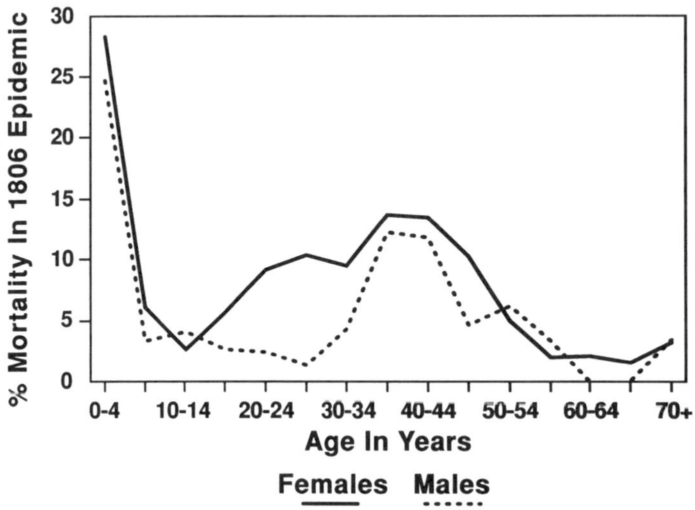

Fig. 5. Age-specific mortality rates for Chumash living at the missions who died during the 1806 measles epidemic. The percent mortality values were calculated for each age group by dividing the number of people in each age group that died during the epidemic by the number of people in same age group who were alive at the end of the month preceding the epidemic. The low percentage of deaths among the elderly suggests exposure to measles in premission times.

It reached a low in the first decade of the nineteenth century, and then rebounded somewhat toward the end of the mission period. For the combined populations of Santa Barbara, Santa Inés, and La Purísima, the average birth rate was 36 per 1,000 people per year. This birth rate is comparable to that of many eighteenth- and nineteenth-century European populations (Ogburn and Nimkoff, 1947:318–321).

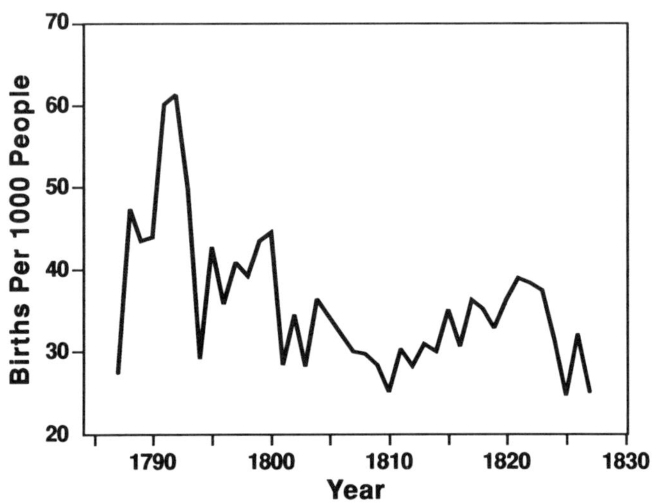

Fig. 6. Birth rate of Chumash living at Mission Santa Barbara, La Purísima, and Santa Inés.

Fig. 7. Changes in the sex ratio of village-born and mission-born Indians living at the Chumash missions.

SEX DIFFERENCES IN MORTALITY

The decrease in birth rate can in large part be explained by the increase in the ratio of men to women noted in our discussion of the 1782 and 1822 Chumash population pyramids (Fig. 2). For a short period at the beginning of the mission period, the mission Chumash sex ratio declined (Fig. 7). This is explained by an excess of young men among the earliest mission converts (Walker and Johnson, 1992). During the ensuing years, these men were followed by women who earlier had remained in the villages. Around 1800, this trend began to reverse. For the rest of the mission period, there was a steady increase in the mission population's sex ratio (Walker and Johnson, 1992).

Historical documents suggest that this sex ratio increase is a result of a high death rate among Chumash women who moved to the missions from native villages (Fig. 7). In a letter written in 1820, Father Mariano Payeras of Mission La Purísima noted that when the Chumash moved to the missions "they decline noticeably in health, get thin, sick and die. This plague is noticed especially in the recently pregnant women" (Payéras, 1820:114). Some deaths may have resulted from attempts to induce abortions. José Longinos Martínez, who visited the Santa Barbara area in 1792, made the following comment on the practice of abortion among the Chumash:

> In this region they have the notion that unless they have an abortion at their first pregnancy, or if the child does not die immediately, they will never conceive again. Hence they murder many babies with the efforts they make, the blows they give themselves, and the barbarous medicine they take in order to induce an abortion, so that some of the women die and others are badly injured. (Simpson, 1961:56)

Sex differences in deaths during the 1806 measles epidemic is consistent with the observations of Father Payéras on the vulnerability of pregnant women living at the missions. This epidemic killed many women in their prime reproductive period between 20 and 34 years (Fig. 5). Mortality among men of the same age was much lower. Considering the comparatively high fertility rate of the Chumash, it is likely that many of these women were pregnant at the time of death. Depression of the immune system during pregnancy provides a possible explanation for the high death rate of young women in this epidemic. In cases of smallpox, for example, pregnancy is known to markedly increase the severity of the disease. This is believed to be a result of elevated levels of hormones that inhibit the immune system (Fenner et al., 1988:38). The possible demographic impact of maternal deaths following childbirth was investigated by determining the number of mothers and fathers who died immediately after the birth of a child. Our assumption was that childbirth-related deaths would result in a large discrepancy in the death rate of mothers and fathers in the year immediately following a birth. For a sample of 987 Mission Santa Barbara births, we found that the death rate of mothers in the year of a birth was higher than that of fathers. The difference, however, is not great; during the year of a birth, 2.8% of mothers and 1.7% of fathers died. During the following year, when childbirth complications are less likely to have caused maternal deaths, 3.8% of mothers and 3.7% of fathers died.

The sex ratio of the mission-born segment of the population counteracts the sex ratio increase discussed above (Fig. 7). These mission-born Chumash made up a small portion of the mission population. The predominance of females among them, therefore, contributes little to the sex ratio of the entire population.

The large number of females in the mission-born Chumash population is of interest because of its possible relationship to unhealthy living conditions. In comparison to females, male development is more easily disrupted by stressful conditions (Stini, 1969; Hamilton, 1982:136; Stinson, 1985). When cultural preferences for male infants do not intervene, this can result in high male mortality rates in unfavorable environments like those that existed at some missions.

UNEVEN SPREAD OF EPIDEMICS

The irregular spatial distribution of Chumash epidemics gives little support to Dobyns's (1983) controversial hypothesis that widespread and devastating epidemics of diseases such as smallpox rapidly spread throughout North America during the sixteenth century. The Chumash missions were located within a short distance from each other, and there was a continual flow of people between them. In spite of this potential for the spread of highly contagious diseases, epidemics sometimes occurred at some missions but not at others. This failure of infections to immediately affect the entire mission population is inconsistent with the reasonably even spread of disease postulated by Dobyns (1983).

The Chumash smallpox experience is particularly revealing in this regard. As early as 1798, there were opportunities for its introduction directly from Europeans. In that year, a ship containing smallpox victims landed at Santa Barbara (Cook, 1939:163–164). Captain Goycoechea of the Santa Barbara Presidio attempted to quarantine the ship's crew, but he was unsuccessful. In spite of this, there was no epidemic among the Chumash. During the next 40 years, smallpox occurred periodically in other parts of California. Yet, none of these epidemics is known to have affected the Chumash. It was not until 1844 that the first cases of smallpox occurred among the Chumash. Even then, it was an isolated phenomenon mainly affecting the northern Chumash missions.

CONCLUSION

Mission records reveal that the Chumash population decline during the nineteenth century was associated with a complex set of demographic changes. Although death rates fluctuated markedly through time, there is a clear, long-term trend toward increasing mortality. A breakdown in mission isolation brought on by an increase in their economic dependence on foreign traders is a likely cause of higher death rates. As the intensity of economic interaction between the missions and the outside world increased, so did opportunities for the introduction of highly contagious infectious diseases that were new to the Chumash. After mission secularization, there was a general deterioration in living conditions. The difficult conditions, along with the spread of chronic diseases such as syphilis and tuberculosis, accelerated the Chumash population decline.

An exceptionally high infant death rate is one striking feature of the mission Chumash mortality profile. Unfortunately for the Chumash, few of their babies survived the first four years of life.

A decrease in births over time in the mission communities also contributed to the population decline. This is explained in part by sex differences in adult mortality. Young women in their prime reproductive years had a higher death rate than men. Such disparities had long-term consequences for the reproductive capacity of the population.

ACKNOWLEDGMENTS

We are indebted to Fr. Virgilio Biasiol of the Santa Barbara Mission Archives and Msr. Francis Weber of the Chancery Archives of the Archdioceses of Los Angeles for providing us with access to the mission records used in this research.

REFERENCES CITED

Bolton HE (1926) Historical Memoirs of New California by Fray Francisco Palóu, O.F.M., vol. 2. Berkeley: University of California Press.

Brown AK (1967) The Aboriginal Population of the Santa Barbara Channel. University of California, Archaeological Survey Report 69. Berkeley: University of California.

Burrus EJ (1967) Diario del Capitán Comandante Fernando de Rivera y Moncada. Colleción Chimalistac de Libros y Documentos Acerca de la Nueva España, vols. 24 and 25. Madrid: Ediciones José Turanzas.

Cook SF (1939) Smallpox in Spanish Mexican California 1770–1845. Bulletin of the History of Medicine 8:151–191.

Cook SF (1976) The Conflict between the California Indian and White Civilization. Berkeley: University of California Press.

Coombs, G (1975) Migration and Adaptation: Indian Missionization in California. Ph.D. dissertation, Department of Anthropology, University of California, Los Angeles.

Costello JG (1990) Variability and Economic Change in the California Missions: An Historical and Archaeology Study. Ph.D. dissertation, Department of Anthropology, University of California, Santa Barbara.

Costello JG (1992) Not peas in a pod: Documenting diversity among the California missions. In BJ Little (ed.): Text-Aided Archaeology. Boca Raton: CRC Press.

Dobyns HF (1983) Their Number Become Thinned: Native American Population Dynamics in Eastern North America. Knoxville: University of Tennessee Press.

Engelhardt Z (1930a) The Missions and Missionaries of California, vol. II, rev. ed. Santa Barbara: Mission Santa Barbara.

Engelhardt Z (1930b) San Buenaventura: The Mission by the Sea. Santa Barbara: Mission Santa Barbara.

Engelhardt Z (1932a) Mission La Concepcion Purísima de Maria Santisima. Santa Barbara: Mission Santa Barbara.

Engelhardt Z (1932b) Mission Santa Inés: Virgen Y Martir and its Ecclesiastical Seminary. Mission Santa Barbara, Santa Barbara, California.

Fenner F, Henderson DA, Arita I, Jezek Z, and Ladnyi LD (1988) Smallpox and its eradication. History of International Public Health 6. Geneva: World Health Organization.

Geiger M, and Meighan C (1976) As the Padres Saw Them. Santa Barbara: Santa Barbara Mission Archive Library.

Hamilton ME (1982) Sexual dimorphism in skeletal samples. In RL Hall (ed.): Sexual Dimorphism in *Homo sapiens:* A Question of Size. New York: Praeger.

Hoover RL (1989) Spanish-native interaction and acculturation in the Alta California missions.

In DH Thomas (ed.): Columbian Consequences, vol. 1, Archaeological and Historical Perspective on the Spanish Borderlands West. Washington, DC: Smithsonian Institution Press, pp. 395–406.

Hornbeck D (1983) California Patterns: A Geographical and Historical Atlas. Palo Alto: Mayfield Publishing.

James GW (1905) In and Out of the Old Missions of California. Boston: Little, Brown and Company.

Johnson JR (1982) An Ethnohistoric Study of the Island Chumash. M.A. thesis, Department of Anthropology, University of California, Santa Barbara.

Johnson JR (1988) Chumash Social Organization: An Ethnohistoric Perspective. Ph.D. dissertation, Department of Anthropology, University of California, Santa Barbara.

Johnson JR (1989) The Chumash and the Missions. In DH Thomas (ed.): Columbian Consequences, vol. 1, Archaeological and Historical Perspective on the Spanish Borderlands West. Washington, DC: Smithsonian Institution Press, pp. 365–376.

Johnson JR (1993) The Chumash Indians After Secularization. In H Benoist (ed.): The Spanish Missionary Heritage of the United States, San Antonio: National Park Service and Los Compadres de San Antonio Missions National Historical Park, in press.

Kenneally F (1965) Writings of Fermin Francisco de Lasuen, vol. 2. Washington, DC: Academy of American Franciscan History.

King CD (1984) Ethnohistoric Background. Appendix 1, Archaeological Investigation of the San Antonio Terrace, Vandenberg Air Force Base, California, vol. 1. Report on file, Department of Army, Corps of Engineers, Los Angeles District.

Lambert PM, and Walker PL (1991) Physical anthropological evidence for the evolution of social complexity in coastal southern California. Antiquity *65*:963–973.

Milliken R (1987) Ethnohistory of the Rumsen. Papers in Northern California Anthropology, 2. Berkeley: Northern California Anthropology Group.

Ogburn WF and Nimkoff MF (1947) A Handbook of Sociology. London: K. Paul, Trench, Trubner and Co.

Ogden A (1975) The California Sea Otter Trade, 1782–1848. Berkeley: University of California Press.

Palóu F (1913) The Life and Apostolic Labors of the Venerable Father Junipero Serra, Founder of the Franciscan Missions of California. Translated by CS Williams. Pasadena: G.W. James.

Payeras M (1820) Letter to Rev. Father and Venerable Director of our Apostolic College of San Fernando, Mexico. Historia de México, Primera Serie, Tomo 2. Mexico City: Archivo General de la Nación.

Reid H (1978) Letters on the Los Angeles County Indians. In SB Dakin (ed.): A Scotch Paisano in Old Los Angeles. Berkeley: University of California Press, pp. 215–286.

Simpson LB (1961) Journal of José Longinos Martínez 1791–1792. San Francisco: John Howell Books.

Stini WA (1969) Nutritional stress and growth: Sex differences in adaptive response. American Journal of Physical Anthropology *31*:417–426.

Stinson S (1985) Sex differences in environmental sensitivity during growth and development. Yearbook of Physical Anthropology *28*:123–147.

Storke Y (1891) A Memorial and Biographical History of Santa Barbara. Chicago: The Lewis Butler Company.

Walker PL (1989) Cranial injuries as evidence of violence in prehistoric southern California. American Journal of Physical Anthropology *80*:313–323.

Walker PL, and Hudson T (1993) Chumash Healing: Changing Health and Medical Practices in an American Indian Society. Banning: Malki Museum Press.

Walker PL, and Johnson JR (1992) Effects of contact on the Chumash Indians. In JW Verano and DH Ubelaker (eds.): Disease and Demography in the Americas. Washington, DC: Smithsonian Institution Press, pp. 127–139.

Walker PL, Lambert P, and DeNiro M (1989) The effects of European contact on the health of California Indians. In DH Thomas (ed.): Columbian Consequences, vol. 1, Archaeological and Historical Perspective on the Spanish Borderlands West. Washington, DC: Smithsonian Institution Press, pp. 349–364.

Chapter 11

Health and Death at Tipu

Mark N. Cohen, Kathleen O'Connor, Marie Danforth, Keith Jacobi, and Carl Armstrong

Department of Anthropology, State University of New York, College at Plattsburgh, Plattsburgh, NY 12901 (M.N.C.,C.A.); Department of Anthropology, University at Albany, State University of New York, Albany, NY 12222 (K.O.); Department of Sociology and Anthropology, University of Southern Mississippi, Hattiesburg, MS 39406 (M.D.); and Department of Anthropology, Indiana University, Bloomington, IN 47405 (K.J.)

The Maya town of Tipu in what is now Belize was the site of a *visita* mission on the fringes of the Spanish colonial empire in the sixteenth century (Jones et al., 1986; Jones, 1989; Graham et al., 1989). The site appears to have been in a volatile no-man's-land between areas of direct Spanish control in Yucatan and parts of the Peten controlled by Itza Maya of Tayasal. Tipu acted as a refuge for Yucatec Maya fleeing more direct Spanish control to the north. It also acted as a center for the cacao trade for much of the Peten.

Recent work at Tipu allows us to utilize historical, osteological,and archaeological data to assess the impact of Spanish influence on the lives of the Tipu Maya.

THE COLONIAL HISTORY OF TIPU

Historical documents suggest that Tipu was inhabited exclusively by Maya. As a *visita* mission, it would have been served by a circuit-riding Spanish priest assisted by local Maya. According to the 1618 census, the community consisted of about 350 people. However, if the colonial documents are accurate, the population appears to have fluctuated dramatically. By 1622, the population may have dropped to as few as 30 people; in the 1640s it rebounded to 1,000 or more. By 1680, the population had dropped to 700, and by 1697 it fell to 400 (Jones, 1989:116).

The Spanish almost certainly had a significant impact on the Tipu community even before first direct contact. Epidemic diseases were introduced to Yucatan as early as 1517 by Spanish soldiers from Cuba. It has been estimated that the population of Yucatan was seriously reduced from at least 800,000 people to as a few as 250,000 before 1544, which is when Spaniards first reached Tipu (Graham et al., 1989). The Yucatan was visited repeatedly by epidemics during the sixteenth and seventeenth centuries: in 1566, 1569–1570, 1575–1576, 1609 (typhus), 1648–1650 (yellow fever, smallpox), 1659 (measles, smallpox), 1692–1693, and 1699 (Farriss, 1984:61). Famines were almost equally common, occurring in 1535–1541, 1564, 1571–1572, 1575–1576, 1604, 1618, 1627–1631, 1650–1653, 1692–1693, and 1700 (Farriss, 1984:61). Similar

In the Wake of Contact: Biological Responses to Conquest, pages 121–133

sequences of events are reported from areas of Guatemala for the same period (Veblen, 1982; Lovell, 1982).

Direct Spanish influence at Tipu began in 1544 following a "violent, vicious" *entrada* (Jones, 1989:22). However, this initial contact did not seem to result in significant reorganization of the Tipu community (Graham et al., 1989). In 1567 and 1568, the Spanish used Tipu as a base for their efforts at pacifying the surrounding country, and their impact on the community was at that time more significant.

In 1638, the Tipu Maya overthrew Spanish control, remaining largely independent for 50 years. In the 1690s, the population at Tipu reestablished contact with the Spanish at Merida, initiating a period of peace and the reestablishment of Christianity in the region. However, after the conquest of the Itza in 1697, Tipu apparently declined in importance. Occupation of the site finally ended in 1707 when Tipu's inhabitants were forcibly removed to the Peten.

ARCHAEOLOGY OF TIPU

Archaeological work at Tipu indicates the presence of some Maya from the late Maya Preclassic (ca. 300 B.C.) through the Historic periods. Tipu was, in fact, a community of some size and importance at the time of initial Spanish contact in 1544. Spanish goods confirm the existence of trade during the period of mission activity. More important, the presence of such goods in typically Postclassic architectural features of pagan rather than Christian style suggests that a non-Christian population—or at least non-Christian activities—extended into the period of Spanish trade.

Trade goods and ceramic styles also suggest that Tipu was part of the Itza economic sphere. Graham and colleagues (1989) note that the Spanish had relatively little impact on Maya material culture at Tipu. Only a veneer of trade items was superimposed on a Maya economy, not the economic disruptions brought about by demands for tribute that occurred at other sites. Maize and squash served as major crops, and the community continued to exploit a wide range of fauna. The comparatively independent Tipu population appears to have fared somewhat better than populations experiencing *reducción*, such as the forcibly relocated population of the neighboring and partly contemporary colonial era population at Lamanai located 100 km north of Tipu; at least Tipu seems to have enjoyed relatively privileged access to Spanish goods (Graham et al., 1989).

The remains of only one Christian church have been found at Tipu. Utilization of the church for the deposition of human remains is clearly bounded by the dates 1544 and 1707, which mark the founding and abandonment of the colonial town, but its use may have been of even shorter duration. It appears that the church does not represent the very earliest period of Spanish influence on the site. Graham and colleagues (1989) believe that the church cemetery probably came into use in 1567–1568, and that the church was desecrated and abandoned during the historically documented rebellion of 1638.

The possibility cannot be ruled out, however, that interment continued after 1638. Graham's excavations have revealed one structure built over a collapsed section of the church wall, a structure which contains intrusive Christian-style burials. The implication is that at least some Christian burial continued after the desecration of the church. Moreover, some burials in the surrounding churchyard seem to have been interred in a manner not properly aligned with the majority of those in the cemetery. These individuals might represent continued burial in once-hallowed ground by people who no longer had a standing church to orient them. Furthermore, there are burials outside the church to the west and north that on the basis of dental discrete traits appear to be separate from the main population (Jacobi,

1993). Such an occurrence would suggest that adherence to the Catholic faith had been sufficiently incorporated into the colonial Maya culture and that it was being maintained without Spanish enforcement.

It is difficult to be certain what portion of the larger Tipu community is represented by the population buried in the church. We do not know whether the mission embraced the whole community or whether some segments of it remained aloof from Christian ritual in life and in death.

THE SKELETAL SAMPLE

Excavations uncovered nearly 600 generally well-preserved skeletons of which 588 represent the Historic period. Of these, 253 are intact primary burials, and an additional 106 are partly disturbed primary burials. The remainder are burials seriously disturbed by subsequent inhumations. Burials are densely packed in the nave of the church, and they extend outside the church on the northern, southern, and western sides of the nave for a distance of 10 to 12 m.

The Tipu burials follow a fairly common pattern of interment. The bodies are supine with their heads to the west and arms folded across the chest or pelvis. The cramped position of the feet and the occasional presence of bone or metal pins on the body suggest that individuals were normally interred in shrouds. Few other goods are found in the graves. One individual (a Maya man of relatively advanced age buried front and center in the nave of the church) was buried in a coffin as indicated by the patterning of nails.

In contrast to some precontact Maya sites, we have found no worked or burned human bone, suggesting that bodies at Tipu were generally treated in a Christian manner after death. In addition, there are few signs of cranial deformation or the filing or inlaying of teeth. Apparently, bodies were treated in Christian fashion during life as well as at death.

Most individuals are buried independently at different levels. There do not appear to be mass burials of the type that might occur after epidemics. Most bodies clearly have been interred with care. The vertical alignment of groups of burials suggests repeated reuse of the space in the nave and the presence of "family plots." Several unusual discrete dental traits cluster together within the cemetery strengthening the family-plot hypothesis (Jacobi, 1993). These traits include labial grooves on central incisors, a deflecting wrinkle on lower molars, sixth and seventh cusps on lower molars, enamel pearls, three-rooted lower molars, odontomes, palatal canines, and peg-shaped teeth. Discrete dental trait analysis also suggests that there was little if any Spanish admixture to the population (Jacobi, 1993).

BIOLOGICAL ANTHROPOLOGY

The quality of skeletal preservation is uneven although excellent by comparison with other Maya samples. Because of differential bone preservation and the mixed nature of a portion of the skeletal material, sample sizes differ among the various analyses.

Demography

For the Historic period, we have identified 173 males, 119 females, 47 adults of unknown sex, and 249 children. Of these 588 burials, 492 are preserved well enough for age-at-death estimates. The skeletal age estimation techniques used and the number of individuals whose ages were estimated by each indicator are shown in Table 1. Summary age estimates were generated for each skeleton by averaging age indicators available for each individual. An age-at-death distribution, constructed from the summary age estimates (Fig. 1), indicates underrepresentation of individuals at the youngest and oldest ends of the life span.

The first two age categories (0–2) are clearly underrepresented. This pattern pre-

TABLE 1. Skeletal Indicators Used to Estimate Age-at-Death for Tipu and the Number of Individuals Available for Each Indicator

	Age indicator	N
Juveniles[a]	Tooth development and eruption[c]	215
	Epiphyseal union[c,d]	95
	Diaphyseal length[c,d]	129
Adults[b]	Auricular surface[e]	128
	Public symphysis[f]	30
	Dental attrition[g]	269
	Cranial suture closure[h]	128
	Cemental annulation[i]	37

[a]Total number of juveniles (≤18 years of age) = 220.
[b]Total number of adults (≥19 years of age) = 272.
[c]Ubelaker (1984).
[d]Bass (1971).
[e]Lovejoy et al. (1985).
[f]Meindl et al. (1985).
[g]Lovejoy (1985).
[h]Stewart (1979); Meindl and Lovejoy (1985).
[i]Data from Wright (1989).

sumably reflects the common archaeological pattern of poor preservation and recovery of the very young. It may also result from the *visita* nature of the Tipu mission. If a priest were only occasionally present, and if unbaptized individuals were denied access to the cemetery, it is likely that many infant deaths would not be among the burials.

Underenumeration of individuals in the oldest age categories is also indicated for this sample. The Tipu age-at-death distribution is an exaggerated example of the disproportionately high numbers of young and mid-adult skeletons, and relatively low numbers of older adults, often observed in osteological samples (cf. Acsádi and Nemeskéri, 1970; Weiss, 1973; Bennett, 1973; Lovejoy et al., 1977; Howell, 1982). Only two individuals 50 years of age or older were identified, representing only 0.4% of the sample.

We have considered four explanations for the apparent absence of older individuals. First, it is possible that there were selective forces at work preventing the burial of elderly individuals who were religiously conservative in a newly established church cemetery (Thwaites, 1959, cited in Devens, 1986:465). However, this possibility only makes sense if the cemetery was of short duration, not the 70–160 years during which the cemetery was used.

Second, documented patterns of frontier immigration, which are likely to have involved disproportionately high percentages of young adults, may have biased the cemetery sample against older individuals. But we would have to assume that the great majority of newcomers also died shortly after their arrival. Otherwise, they too would grow old and add their numbers to the upper end of the age distribution.

Third, methodological problems stemming from the identification of older adult remains may have contributed to the observed pattern. Poor preservation and recovery of elderly remains (see Walker et al., 1988) and error toward age underestimation for some adult aging methods may lead to

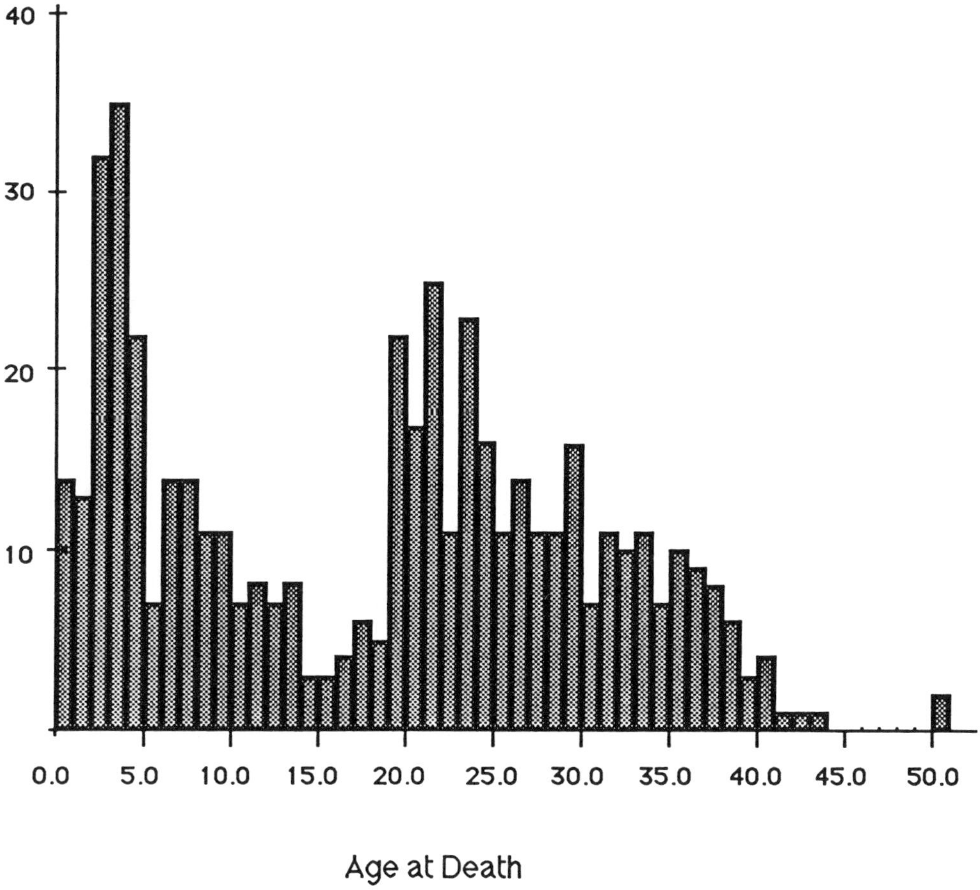

Fig. 1. Age-at-death distribution for Tipu. $N = 492$.

an apparent deficit of older adults and an erroneously high level of mortality in the early and mid-adult ages (Buikstra and Konigsberg, 1985). To evaluate the potential bias introduced by age estimating methods, or by individuals applying these methods, the techniques used for Tipu were applied by the same investigators to skeletal samples from precontact Maya populations at Copan, Barton Ramie, Seibal, and Tikal. We found older adults to be more common in these samples, suggesting that methodological considerations alone do not account for the unusual pattern at Tipu. In addition, Wright (1989) used cemental annulation to estimate the ages of 38 individuals previously aged by macroscopic methods. She produced a mean age-at-death estimate for these 38 individuals which was similar to that derived for the same individuals by our other techniques.

Fourth, it is possible that people died in their prime from epidemic diseases that left no direct mark on the skeleton. However, there is no evidence of hasty or mass burial that might be expected to occur under epidemic conditions. On the other hand, successive waves of different high-mortality diseases might have contributed to the low frequency of older adults. If the younger age cohorts of the population were virgin to each high mortality disease, and epidemics occurred at relatively short intervals (i.e., shorter than a generation), then individuals who survived an epidemic of one disease would subsequently be exposed to one or more additional high-mortality diseases in their remaining lifetimes, leaving few individuals to attain old age. Although no adequate historical data exist to investigate this possibility for the Tipu population, sources do indicate that in some Maya areas epidemics of different diseases occurred approximately every 10 to 20 years (e.g., in highland Guatemala: Veblen, 1982:97; Lovell, 1982: 114; in colonial Yucatan: Farriss, 1984:61).

In addition to the epidemic diseases, another severely debilitating disease, malaria, is believed to have been endemic in the frontier lowland areas, and it may have been an important cause of early adult death (Jones, 1989). Endemic malaria, successive exposure to different high-mortality epidemic diseases, and frequent famines, sometimes the consequence of socioeconomic disruptions following epidemics (e.g., Jones, 1989:334), may have together presented health stresses so burdensome that few individuals could survive them all and, hence, live to old age.

Trauma and Infection

A total of 457 skeletons have been examined for trauma and infectious lesions (Armstrong, 1989; Cohen et al., 1989). Trauma is rare, occurring on less than 2% of individuals, and is most common on tibiae. For adult bones, 13 of 710 tibiae (1.8%) and seven of 610 fibulae (1.1%) show signs of injury, but only one of 649 (0.2%) humeri displays trauma. Among juvenile bones, only one of 236 tibiae examined (0.4%) displays trauma; no other juvenile long bones display trauma. Among whole skeletons, 24 of 275 adults (8.7%) display some trauma. This includes 16 of 154 adult male individuals (10.4%) and eight of 116 adult female individuals (6.9%). In addition, one of 182 juveniles (0.5%) shows trauma.

Most fractures appear accidental, and the frequencies of trauma increase gradually with age, consistent with lifelong exposure to accidental injuries. In addition, one adult has a crushed facial bone (left zygomatic arch) that may reflect violence but which displays signs of healing, one has a probable partially healed sword cut on the frontal bone, another has an entry wound on a tibia that appears to have healed prior to death, and one child has multiple cranial perforations. With these possible exceptions, no bones suggest violence that might have been inflicted by Spanish weapons. Overall, the Tipu pattern of trauma suggests occasional violence, not continuing political strife.

Periosteal reactions of long bones, usually identified with infection, are also infrequent, although they are somewhat more common than trauma. These lesions occur most commonly on the tibia, followed in order by the fibula, femur, radius, ulna, and humerus (Armstrong, 1989). Of 704 tibial diaphyses, 59 (8.4%) display some signs of periostitis, and 33 of 601 fibular diaphyses (5.5%) are similarly affected. Other portions of these bones, as well as other long bones, have lesion frequencies of 1.0% or less. Long bones of males were slightly more likely to display signs of infection than female bones. Only five of 236 juvenile tibiae (2.1%) and two of 208 juvenile fibulae (1.0%) have these lesions.

Among whole individuals, 36 of 159 males (22.6%) display a pathological involvement of at least one bone; 16 of 116 females (13.8%) are similarly afflicted. Four of 182 juveniles (2.2%) show signs of these lesions.

Examples of systemic infection are found in 9.9% of males, 7.5% of females, and only 1.1% of juveniles (i.e., lesions on two or more limbs, noncontiguous limb segments, or on other parts of the skeleton assuming the lesions are related to one another). Overall, 19 of 457 individuals (4.2%) display systemic infection. Most changes in bone are of a relatively minor order. Only two cases represent a more severe involvement and deep penetration of bone (osteomyelitis), and both cases are associated with trauma.

Although tibiae are the most commonly affected bones, the slight nature of most lesions, their unilateral character, and the absence of lesions on other bones such as the cranium suggest that treponemal infection was not present in the Tipu population. Furthermore, there are no signs of the dental stigmata characteristic of congenital treponemal infection (Jacobi et al., 1992). Although the preserved sample is much smaller, the absence of collapse of lumbar vertebrae and the absence of rib lesions suggest that skeletal tuberculosis or tuberculosis-like infections did not occur at Tipu.

The proportion of skeletons with skeletal lesions, primarily periostitis, is low by comparison with most prehistoric populations (see Cohen and Armelagos, 1984; Cohen, 1989). It does not suggest the presence of widespread chronic illness which we expected to find. The meaning of such low figures is not clear. The church cemetery may represent a privileged segment of the Tipu community (Armstrong, 1989). Alternatively, it may reflect a real infrequency of chronic infection, or it may result from the rapid death of individuals from acute introduced diseases instead of from indigenous chronic illnesses that affected bone (see Wood et al., 1992).

Enamel Hypoplasia and Wilson Bands

A sample of 180 dentitions has been examined for linear enamel hypoplasia on upper central incisors and mandibular canines (Cohen et al., 1989). Teeth were examined with a $\times 7-\times 30$ binocular dissecting microscope so that lesions not usually visible to the naked eye could be recorded (see Goodman et al., 1980). The position of each lesion was identified by measuring the distance from its center to the cementoenamel junction, and ages of defects were estimated following the method of Goodman and colleagues (1980). Sixty-nine percent of incisors and 90.4% of canines had at least one hypoplasia; 100 of 105 individuals (95%) observable for both teeth had at least one such defect on one or both teeth. However, 88% of enamel defects recorded on incisors were mild. Severe enamel hypoplasia occurred on only 2.2% of incisors and 10.3% of canines. Only two of 105 individuals (1.9%) with both teeth had severe lesions on both teeth. The distribution of 195 hypoplastic defects on incisors suggests a peak age of formation at about 2.5–3 years. Examples of enamel hypoplasia ($n = 278$) on the canine also display a mild peak at this age, but the greatest number of defects developed between 4 and 4.5 years of age. Ethnohistoric references (Landa, 1978) report that Maya weaned their children at about ages three to four, and this corresponds roughly to the peaks of our enamel hypoplasia data.

Only 30 individuals in this sample had three or more separate enamel defects on canines. Of these, only 12 display regular spacing. The pattern suggests that regularly occurring stresses account for only a small part of the health problems experienced by people at Tipu.

A comparative study of macrodental and microdental defects in a sample from the Tipu population ($N = 191$) and a series of precontact, Late Classic Maya samples from Tikal ($N = 28$), Barton Ramie ($N = 37$), and Seibal ($N = 27$) was also undertaken (Danforth, 1989). Three types of dental defects were scored, including striae of Retzius, Wilson bands, and enamel hypoplasia; these defects represent different forms of interruption to dental development (Hillson, 1986: 107–150). The relatively high frequency of striae in most teeth suggests that they resulted from mild stress. Wilson bands, which are striae with a disruption in prism orientation, are associated with short-term acute stress, whereas enamel hypoplasias are usually associated with longer-term health disruption (Rose et al., 1985).

Defects were observed by viewing longitudinal sections of deciduous canines, permanent canines, and third molars under magnification (×40–×400). Danforth (1989) concluded that the postcontact Tipu population did not display patterns of enamel defects suggesting high levels of stress associated with epidemic or enzootic infections or with severe nutritional deprivation. Instead, the pattern of dental defects is consistent with the image of a generally healthy, even privileged, population. Moreover, the Tipu sample displayed fewer dental defects than the Late Classic samples. While the Tipu sample had high frequencies of striae of Retzius suggesting relatively frequent mild stresses, the Late Classic samples displayed a higher frequency and greater severity of the more severe defects (Wilson bands and severe enamel hypoplasia). Both groups average about six striae per canine, but the Late Classic populations had significantly more Wilson bands. Similarly, the earlier Maya averaged 3.75 enamel hypoplasias per canine as compared to 2.75 in the Historic period sample. In addition, hypoplastic defects in the Late Classic group were more than twice as likely to be classified as severe. Finally, all three defects occurred together most often in the Late Classic groups.

Danforth (1989) also found that both the Late Classic and Historic Maya samples displayed relatively few disruptions in tooth development through the first two postnatal years of life. Evidence for prenatal stress was very rare, suggesting that these populations did not suffer a high degree of severe maternal malnutrition (Danforth, 1989; Cohen et al., 1989; see Blakey and Armelagos, 1985). There is some increase in the frequency of stress during the perinatal period, but early childhood is then a time of little stress. Only 26% of deciduous canines in the Tipu sample show enamel hypoplasias when viewed microscopically, and most of these are shallow, short, and probably not observable by macroscopic evaluation. Observations of permanent canines also suggest that stress was rare for individuals under two years of age, but the samples show an increase in enamel defects related to the probable age of weaning between three and five years. Neither the Colonial nor the Classic dentitions commonly display signs of annual stress (Danforth, 1989).

Bone Cortical Area

Femoral cortical involution in a sample of 127 adults individuals and 37 children under 12 years of age has been studied by Danforth and colleagues (1985). Each femur was sectioned immediately distal to midshaft and cortical area was measured. In addition, the thickness of the cortex was measured at 12 points around its circumference. Mean cortical thickness was expressed as a function of femur length, and Nordin's index (Nordin, 1962) and percent cortical area were calculated. Osteoporosis was found in only 3% of adult individuals after adjusting Nordin's radiographic standard for dry bone. This quality of bone maintenance may be related to good calcium/phosphorous ratios in the diet resulting from maize preparation methods or grit from limestone metates (Katz et al., 1974; Coe, 1980:599–600; Cotron, 1989). However, it should be noted that the sample is composed primarily of young adults, which makes finding osteoporosis unlikely. There was a lack of sexual dimorphism in Nordin's index resulting from a low absolute level of cortical bone in men, perhaps suggestive of some malnutrition.

Danforth and coworkers (1985) also reported that cortical bone area and femoral diaphyseal length increased regularly with age in their small sample of juvenile individuals, a result later confirmed by an analysis of 96 juvenile skeletons from the entire population (Bennett and Casey in Cohen et

TABLE 2. Mean Stature (in cm) by Sex at Selected Maya Sites

Site	Males (n)	Females (n)
Tipu	160.3 (149)	148.3 (106)
Tancah[a]	157.9 (4)	—
Altar de Sacrificios[b]		
Preclassic	166.6 (4)	148.3 (2)
Early Classic	159.4 (2)	149.5 (1)
Late Classic	162.0 (1)	—
Postclassic	—	146.5 (1)
Tikal[c]		
Preclassic	161.7 (6)	144.4 (4)
Early Classic	164.2 (15)	139.4 (3)
Late Classic	155.2 (21)	148.5 (11)
Postclassic	—	148.5 (1)
Barton Ramie[d]		
Late Classic	156.6 (10)	145.7 (6)
Copan[d]		
Late Classic	138.0 (11)	147.2 (17)
Modern Maya[e]	155.2 (128)	142.8 (94)

[a]Saul (1982).
[b]Saul (1972).
[c]Haviland (1967).
[d]Cohen et al. (1989).
[e]Steggerda (1941).

al., 1989). Tipu children showed slight but steady gains in bone thickness through childhood, suggesting that severe protein-calorie malnutrition was not present (Garn et al., 1966; Huss-Ashmore et al., 1982).

Stature

Adult statures at Tipu are at least average by available Maya data. They do not appear to indicate that growth was compromised by significant economic deprivation relative to precontact populations. One hundred and forty-nine adult male individuals in the Tipu sample have an average estimated stature of 160.3 cm (formulas in Genovés, 1967); 106 adult females have an average stature of 148.3 cm. Mean statures from several other Maya sites are given in Table 2. All stature estimates have been calculated based on published data using the same formulas (Genovés, 1967). The values for Tipu are well within the ranges of earlier sites in the region.

Anemia

Cribra orbitalia and porotic hyperostosis, both symptoms of anemia, have been evaluated by several workers. Of 214 individuals with one or two intact orbits, cribra orbitalia occurs in 39 (18.2%). Cribra orbitalia occurs in 8.8% of adult men (8 of 91), in 5.4% of adult women (3 of 56), and in 30.5% of children (18 of 59).

Porotic hyperostosis was scored in 304 individuals with sufficiently intact skulls. Of these, 16 (5.3%) display active lesions, and 61 (20.1%) display healed lesions. Among 116 males scored, 29 individuals (25%) were positive. Among adult women, eight of 69 (11.6%) were positive. The difference between the sexes was found to be statistically significant (χ^2, $p < 0.05$). Among children, 15 of 106 (14.2%) display active porotic hyperostosis; 23 of 106 (21.7%) display healed or healing porosity. Active anemia was found in 27% (10 of 37) of children dying at ages 9 or older, but in only five of 69 chil-

TABLE 3. Proportion of Individuals with Porotic Hyperostosis by Site

Site	Males (n) (healed)	Females (n) (healed)	Juveniles (n) (healed and active)
Tancah[a]	0.833 (6)	0.000 (1)	1.000 (3)
Chichen Itza[b]	0.600 (10)	0.429 (7)	0.778 (18)
Lamanai[c]	0.696 (23)	0.640 (25)	0.800 (10)
Tipu	0.241 (116)	0.116 (69)	0.358 (106)
Altar de Sacrificios[d]	0.857 (14)	0.923 (13)	0.857 (7)
Seibal[e]	0.286 (14)	0.428 (7)	0.333 (1)

[a]Saul (1982).
[b]Hooton (1940).
[c]White (1986).
[d]Saul (1972).
[e]Cohen et al. (1989).

dren (7.2%) of those dying at ages 8 and younger. Similarly, remodeled lesions were found in 15 of 37 children dying at age 9 and older (40.5%) but in only eight of 69 children (11.6%) dying at younger ages.

These figures are low relative to figures published from other Maya sites, but similar to frequencies of anemia that we found at Barton Ramie, Seibal, and Copan (Table 3). The differences in scoring methods among researchers may account for some variation. On the other hand, few Tipu crania display lesions of the severity seen in published photographs of skulls from Altar de Sacrificios (Saul, 1972).

CONCLUSION

Reconstruction of the effects of European contact on the health and culture of Native American populations is complicated by the various methodological and analytical problems associated with osteological data from archaeological contexts. In particular, the levels and age patterns of both mortality and morbidity extrapolated from skeletons can be influenced by nonrepresentative samples (due to cultural or taphonomic factors), competing causes of death, and problems associated with age-at-death estimation methods (Moore et al., 1975; Buikstra and Konigsberg, 1985; Wood et al., 1992). Therefore, inferences about past health based on such data remain tentative. However, supporting evidence from other sources, such as ethnohistoric documents, often can be brought to bear in an attempt to tease out the various factors contributing to the osteological picture of lesions and death. For Tipu, inferences about past health from skeletons can be supplemented by ethnohistoric data that paint a picture of intermittent conflict, famine, and epidemic disease.

The Tipu cemetery represents some portion of a Christian Maya population in periodic contact with the Spanish between 1544 and 1707. The sample therefore postdates the first waves of epidemics in the Yucatan, but it was contemporary with significant outbreaks of new diseases, famine, and political unrest.

Although a seemingly straightforward interpretation of the Tipu osteological data would appear to indicate relatively good health, as evidenced by the low rates of pathology (i.e., infection, enamel hypoplasia, anemia, trauma), the demographic and historic data raise the strong possibility that epidemics of new, acute infectious disease epidemics took a heavy toll on Tipu's inhabitants.

The Tipu sample displays a marked scarcity of signs of violent trauma, despite periodic tensions with the Spanish, and the suspected refugee status of some of Tipu's inhabitants who presumably fled from areas of more sustained Spanish control. The osteological data thus suggest that the people at Tipu were relatively insulated from direct conflict.

The sample also displays levels of skeletal infection which are low, or moderate, by the standards of neighboring or precolonial Maya groups, or of prehistoric samples generally (see various studies in Cohen and Armelagos, 1984). At first glance, such a pattern seems at odds with the site's historical setting. However, lower than expected frequencies of pathology may be, in part, a result of many rapid deaths from acute epidemic diseases. The existence of new, high-mortality infectious diseases, in turn, may have played a role in the observed degree of adherence to Christian-introduced lifeways and may help explain the care with which individuals were interred. Christian-style burial as well as the lack of body decoration or deformation appears to have been the rule for Tipu even though the site was located at the remote frontier of Spanish influence. Such adherence to Christian doctrine may have been facilitated by Native Americans' observations of the relative immunity of Christian Europeans to epidemic diseases compared to the repeated and heavy losses they themselves suffered (McNeill, 1976).

There are, however, alternative or additional factors that may have contributed to the composition and nature of the Tipu cemetery. Selective cultural forces in both life and interment may have contributed to a cemetery composed primarily of young people who were subjected to relatively few illnesses affecting bone. Additionally, possible underaging of adults and certain underenumeration of infants contribute to what appears as high mortality at the young and middle adult ages.

ACKNOWLEDGMENTS

The excavation and analysis of Tipu have been supported by National Science Foundation grants BNS 83-03693 and BNS 85-06785.

REFERENCES CITED

Acsádi G, and Nemeskéri J (1970) History of Human Life Span and Mortality. Budapest: Akadémiai Kiadó.

Armstrong C (1989) Pathological Analysis of Long Bones Within the Tipu Population. M.A. thesis, Liberal Studies, College at Plattsburgh, State University of New York.

Bass, WM (1971) Human Osteology. Columbia: Missouri Archaeological Society.

Bennett KA (1973) On the estimation of some demographic characteristics of a prehistoric population from the American Southwest. American Journal of Physical Anthropology 39:223–232.

Blakey M, and Armelagos GJ (1985) Deciduous enamel defects in prehistoric Americans from Dickson Mounds. American Journal of Physical Anthropology 66:371–380.

Buikstra JE, and Konigsberg LW (1985) Paleodemography: Critiques and controversies. American Anthropologist 87:316–333.

Coe WR (1980) Artifacts of the Maya lowlands. In GR Willey (ed.): Archaeology of Southern Mesoamerica, Part II. Handbook of Middle American Indians, vol. 3. Austin: University of Texas Press, pp. 594–602.

Cohen MN (1989) Health and the Rise of Civilization. New Haven: Yale University Press.

Cohen MN, and Armelagos GJ (eds.) (1984) Paleopathology at the Origins of Agriculture. Orlando: Academic Press.

Cohen MN, Bennett S, and Armstrong C (1989) Final report to the National Science Foundation on grant BNS-85-06785. Health and genetic relationships in a Colonial Maya population.

Cotron, RS, Kumar V, and Robbins SL (1989) Robbins' Pathological Basis of Disease. 4th ed. Philadelphia: Saunders.

Danforth M (1989) Comparison of Health Patterns in Late Classic and Colonial Maya Populations Using Enamel Microdefects. Ph.D. dissertation, Department of Anthropology, Indiana University, Bloomington.

Danforth M, Bennett S, Cohen MN, and Melkunas H (1985) Femoral cortical involution in a Colonial Maya population. Paper presented, American Association of Physical Anthropologists, Knoxville.

Devens C (1986) Separate confrontations: Gender as a factor in Indian adaptation to European colonization in New France. American Quarterly *88*:461–480.

Farriss N (1984) Maya Society Under Colonial Rule. Princeton: Princeton University Press.

Garn SM, Rohmann CG, and Guzman MA (1966) Malnutrition and skeletal development in preschool children. In: Preschool Child Malnutrition. Washington, DC: National Academy of Sciences-National Research Council, pp. 43–62.

Genovés S (1967) Proportionality of long bones and their relation to stature among Mesoamericans. American Journal of Physical Anthropology *26*:67–77.

Goodman A, Armelagos GJ, and Rose J (1980) Enamel hypoplasia as an indicator of stress in three prehistoric populations from Illinois. Human Biology *52*:515–518.

Graham E, Pendergast DM, and Jones GD (1989) On the fringes of conquest: Maya-Spanish contact in Colonial Belize. Science *246*:1254–1259.

Haviland W (1967) Stature at Tikal. American Antiquity *35*:316–325.

Hillson S (1986) Teeth. Cambridge: Cambridge University Press.

Hooton EA (1940) Skeleton from the Cenote of Sacrifice at Chichen Itza. In CL Hay (ed.): The Maya and Their Neighbors. New York: Dover, pp. 270–280.

Howell N (1982) Village composition implied by a paleodemographic life table: The Libben Site, Ohio. American Journal of Physical Anthropology *59*:263–270.

Huss-Ashmore R, Goodman AH, and Armelagos GJ (1982) Nutritional inference from Paleopathology. In M Schiffer (ed.): Advances in Archaeological Method and Theory, vol. 5. New York: Academic Press, pp. 395–474.

Jacobi K (1993) An Analysis of Genetic Structuring in a Colonial Maya Cemetery, Tipu, Belize, Using Dental Morphology and Metrics. Ph.D. Dissertation, Department of Anthropology, Indiana University, Bloomington, in preparation.

Jacobi K, Cook DC, Corruccini R, and Handler JS (1992) Congenital syphilis in the past: Slaves at Newton Plantation, Barbados, West Indies. American Journal of Physical Anthropology *89*: 145–58.

Jones G (1989) Maya Resistance to Spanish Rule. Albuquerque: University of New Mexico Press.

Jones G, Kautz R, and Graham E (1986) Tipu: A Maya town on the Spanish colonial frontier. Archaeology *39*:40–47.

Katz SH, Hediger MI, and Valleroy LA (1974) Traditional maize processing techniques in the New World. Science 184:765–768.

Landa D (1978) Yucatan Before and After Conquest. New York: Dover.

Lovejoy CO (1985) Dental wear in the Libben population: Its functional pattern and role in the determination of adult skeletal age at death. American Journal of Physical Anthropology *68*:47–56.

Lovejoy CO, Meindl RS, Pryzbeck TR, Barton TJ, Heiple KG, and Kotting D (1977) Paleodemography of the Libben site, Ottawa County, Ohio. Science *198*:291–293.

Lovejoy CO, Meindl RS, Pryzbeck TR, and Mensforth RP (1985) Chronological metamorphosis of the auricular surface of the ilium: A new method for the determination of adult age at death. American Journal of Physical Anthropology *68*:15–28.

Lovell WG (1982) Collapse and recovery: A demographic profile of the Cuchumatan Highlands of Guatemala (1520–1821). In RM Carmack, J Early, and C Lutz (eds.): The Historical Demography of Highland Guatemala. Institute for Mesoamerican Studies, State University of New York, Albany, Publication 6, pp. 103–120.

McNeill W (1976) Plagues and Peoples. Garden City: Anchor.

Megged A (1991) Accommodation and resistance of elites in transition: The case of Chiapa in early Colonial Mesoamerica. Hispanic American Historical Review *71*:477–450.

Meindl RS, and Lovejoy CO (1985) Ectocranial suture closure: A revised method for the determination of skeletal age at death based on the lateral anterior sutures. American Journal of Physical Anthropology *68*:57–66.

Meindl RS, Lovejoy CO, Mensforth RP, and Walker RA (1985) A revised method for age determination using the os pubis with a review and tests of accuracy of other current methods of pubic symphyseal aging. American Journal of Physical Anthropology *68*:29–45.

Moore JA, Swedlund AC, and Armelagos GJ (1975) The use of life tables in paleodemography. American Antiquity *40*:57–70.

Nordin BEC (1962) The problem of osteoporosis. Gerontologia Clinica 4:19–32.

Rose JC, Condon KW, and Goodman AH (1985) Diet and dentition, developmental disturbances. In

RI Gilbert and Mielke JH (eds.): The Analysis of Prehistoric Diets. Orlando: Academic Press, pp. 281–305.

Saul F (1972) The Human Skeletal Remains of Altar de Sacrificios. Papers of the Peabody Museum of Archaeology and Ethnology, Harvard University, 63(2).

Saul F (1982) Appendix II. The human skeletal remains from Tancah, Mexico. In A Miller (ed.): On the Edge of the Sea: Mural Painting at Tancah-Tulum, Quintana Roo, Mexico. Washington DC Dumbarton Oaks, pp. 115–128.

Steggerda M (1941) Maya Indians of Yucatan. Carnegie Institution of Washington, Publication 531.

Stewart TD (1979) Essentials of Forensic Anthropology. Springfield: Charles C Thomas.

Ubelaker DH (1984) Human Skeletal Remains: Excavation, Analysis, Interpretation. Washington, DC: Taraxacum.

Veblen TT (1982) Native population decline in Totonicapan, Guatemala. In RM Carmack, J Early, and C Lutz (eds.): The Historical Demography of Highland Guatemala. Institute for Mesoamerican Studies, State University of New York, Albany, Publication 6, pp. 81–102.

Walker PL, Johnson JR, and Lambert PM (1988) Age and sex biases in the preservation of human skeletal remains. American Journal of Physical Anthropology *76*:183–188.

Weiss K (1973) Demographic Models for Anthropology. Memoirs of the Society for American Archaeology 27.

White C (1986) Paleodiet and Nutrition of the Ancient Maya at Lamanai, Belize: A Study of Trace Elements, Stable Isotopes, Nutrition and Dental Pathology. M.A. thesis, Department of Anthropology, Trent University, Peterborough.

Wood JW, Milner GR, Harpending HC, and Weiss KM (1992) The osteological paradox: Problems of inferring prehistoric health from skeletal samples. Current Anthropology *33*:343–370.

Wright LE (1989) Cemental Annulation: A Test of Its Application to the Colonial Maya of Tipu, Belize. M.A. thesis, Department of Anthropology, University of Chicago.

Chapter 12

Biological Disruption in the Early Colonial Period at Lamanai

Christine D. White, Lori E. Wright, and David M. Pendergast

Department of Anthropology, University of Western Ontario, London, Ontario N6A 5C2, Canada (C.D.W.); Department of Anthropology, University of Chicago, Chicago, IL 60637 (L.E.W.); and Department of New World Archaeology, Royal Ontario Museum, Toronto, Ontario M5S 2C6, Canada (D.M.P.)

Within the 60 years following the arrival of the Spanish in 1519 and, later, their African slaves, records indicate that native Central American populations were dramatically reduced by disease (Thompson, 1966; Crosby, 1967, 1972; Gerhard, 1979; Lovell, 1985). For the Maya region in particular, new epidemics included smallpox (1520), measles (1529), and influenza (1557) (Farriss, 1984). Unfortunately, bioarchaeological research on the effects of conquest has been limited, especially for the Maya. The poor preservation of skeletons in tropical environments, the failure of many infectious diseases to leave characteristic skeletal lesions, and a limited number of excavated sites dating to the transition period has left little with which to work.

The Maya lowlands present an interesting context for studying the consequences of Old World and New World interactions. Maya populations in the region seem to have experienced remarkable cultural stability through the period of conquest (Chase and Rice, 1985; Pendergast, 1986b). With only a few exceptions, evidence for the Spanish presence prior to the nineteenth century is difficult to identify with archaeological materials alone (Chase and Rice, 1985). The most profound changes experienced in this area may have been in demography and health. Patterns of diet and dental and skeletal pathology from the lowland site of Lamanai in northern Belize indicate that the cultural stability that typifies the Postclassic to Historic period transition is overlaid with biological change.

ARCHAEOLOGICAL CONTEXT

Excavation of the ancient Maya site of Lamanai revealed extensive evidence of an apparently continuous occupation from about 1500 B.C. through the Postclassic period (ca. A.D. 950–1544) and into the first century of the Spanish colonial era (Pendergast, 1981, 1986a). Although documentary evidence bearing on the period of Spanish hegemony is extremely limited, it provides a terminal date of 1641, the year in which the site's inhabitants joined an uprising fomented by other communities, destroyed Spanish churches and related buildings, and supposedly fled to the neighboring site of Tipu located further inland (López de Cogolludo, 1971). Archaeological evidence, however, shows that at least

In the Wake of Contact: Biological Responses to Conquest, pages 135–145

part of the populace remained at the site for some time after the last Spanish contact, but burials related to the post-1641 occupation were not revealed in excavations.

The site, situated at the western edge of a large lagoon or lake, was strategically located to take advantage of the avenue of transport provided by the New River, the sixteenth-century *Dzuluinicob* (Jones, 1989), that flows northward into the Caribbean. Together with what were probably many cultural and environmental factors (Pendergast, 1986b), Lamanai's location very likely played a significant role in the community's survival from the social disintegration that marked the end of large-scale construction and dense population at most southern lowlands centers. Though paralleled in part by data from other sites in Belize, including Santa Rita (Chase and Chase, 1988), Tipu (Graham et al., 1985; Graham, 1991; Cohen et al., this volume), and Marco Gonzalez (Graham and Pendergast, 1989; Pendergast, 1991a), the record of Postclassic through Historic occupation at Lamanai remains unusual in lowland Maya prehistory.

The impact of Spanish presence on Maya life at Lamanai is difficult to assess using the limited ethnohistorical documentation, but it is discernible archaeologically (Graham et al., 1989). Furthermore, excavated evidence for the centuries of Postclassic life that preceded the coming of the Spaniards is abundant enough to permit a reasonably gap-free portrayal of the community's cultural development from the mid-tenth through the mid-sixteenth centuries.

The passage of Lamanai through times of crisis in the southern lowlands ultimately resulted in abandonment of most or all civic structures in the Classic site center and a concentration of population near the central precinct's southern limit (Pendergast, 1986b). Despite the collapse of neighboring communities, the people of Lamanai sustained a relatively high level of wealth throughout the centuries that followed, with an economic base that united local productivity with import and export ties to settlements in the northern part of the Yucatan Peninsula. By the time of Spanish arrival in 1544, or shortly thereafter (Jones, 1989), the community was surely much smaller than that of the Classic years, but nevertheless remained vigorous and stable (Pendergast, 1991b). The lone documentary reference to the number of Lamanai's inhabitants seems to fix the settlement at about 72 individuals in 1637 (Jones, 1989), but the date and the events of the contact period make the figure irrelevant to calculations of population size prior to, and immediately following, first European contact. Contrary to expectations based on events in many other parts of the New World, Spanish control over Lamanai had relatively little effect on indigenous material culture (Pendergast, 1991b; Pendergast and Graham, 1991), mainly because their presence was small and interrupted (Graham et al., 1989).

THE SKELETAL SAMPLE

Because Lamanai has not only yielded one of the larger and more continuous lowland Maya samples to date, its human remains are an important representation of Maya biological history. The bulk of Lamanai burials come from the Postclassic ($n = 115$) and Historic ($n = 179$) periods. The utility of the Postclassic burial data are, however, limited because they are the product of a sampling procedure unrelated to patterns of ancient interment and, hence, cannot be clearly defined as a sample of deaths from a real population. It is in this one regard that the contact period evidence is unquestionably superior to that from earlier centuries. Differential Maya response to Christian precepts aside, the 1544–1641 burial sample clearly can be approached as representative of community deaths.

The elite status of the Postclassic burials is demonstrated by their derivation from various ceremonial core contexts. During the

TABLE 1. Age and Sex Representation by Time Period

	Postclassic (*n*)	Historic (*n*)
Sex		
Male	36	41
Female	32	53
Sex ratio	1.13:1	0.77:1
Age		
Adult	68	94
Juvenile[a]	30	55
Age ratio	2.26:1	1.71:1

[a]Less than 20 years.

Historic period, burial still took place in the core, but the sample comes from beneath the floor of the earlier of the community's two churches. The sex ratio during the Postclassic is slightly weighted toward males (1.13:1, M:F), a situation which is reversed in the Historic (0.77:1, M:F) (Table 1). Nonetheless, these differences are not statistically significant (χ^2, $p > 0.05$). There does, however, appear to be a greater representation of juveniles in the Historic sample (χ^2, $p \leq 0.05$).

DIETARY RECONSTRUCTION

Stable Isotopes

The inhabitants of Lamanai had access to foods from diverse ecological zones, both aquatic and terrestrial (Fig. 1). The immediate and surrounding areas include rainforest, savannah, swamp, floodplain, lagoon, riverine, estuarine, and possibly ocean reef. Archaeological data indicate that maize was grown intensively in northern Belize (Harrison and Turner, 1978), as well as at Lamanai (Richard Hebda, personal communication, 1979). Maya historical and ethnohistorical documents are dominated by references to maize as a staple (Béhar, 1968). To determine the relative proportions of maize and other protein sources in the diet, stable carbon and nitrogen isotope ratios were measured for bone collagen (White, 1986; White and Schwarcz, 1989). The $^{13}C/^{12}C$ ratios indicate the quantity of C_4 plants and C_4 plant-consuming animals in the diet. Maize is the only agricultural C_4 plant used by the Maya in Belize, although they could have consumed animals such as dogs, wild deer, and peccary that might have fed on maize. The isotopic values of some of these animals suggest C_4 feeding. Although $\delta^{13}C$ values indicate marked temporal change for periods preceding the Postclassic, there is no change in the size of the C_4 component of the diet from Postclassic to Historic times (*t* test, $p > 0.05$) (Table 2, Fig. 2). The quantity of maize consumed is estimated to have been high (White and Schwarcz, 1989), as are ethnographic estimates of 65% to 86% maize consumption in modern Maya diets (Benedict and Steggerda, 1936).

The balance of protein resources, as indicated by $\delta^{15}N$ values, similarly does not change over time (*t* test, $p > 0.05$, Table 2). The predominant protein source appears to have been terrestrial herbivores, with only marginal consumption of fish and other aquatic resources of high trophic level (White and Schwarcz, 1989). Turtle remains are found in abundance at the site and are characterized as C_3. Molluscs are also a common component of the faunal assemblage, but research by Emery (1986) suggests that they were harvested only at certain times of the year.

Dental Pathology

Because teeth are in frequent and direct contact with food, they are useful dietary indicators. Dental caries and calculus have been measured for all observable Postclassic remains and specimens from one Historic cemetery (N12-11). Tooth decay has repeatedly been associated with high levels of carbohydrates in the diet (see Larsen, 1987). Carious lesions were calculated per individual as a percent of the observable number of teeth. Calculus was scored for

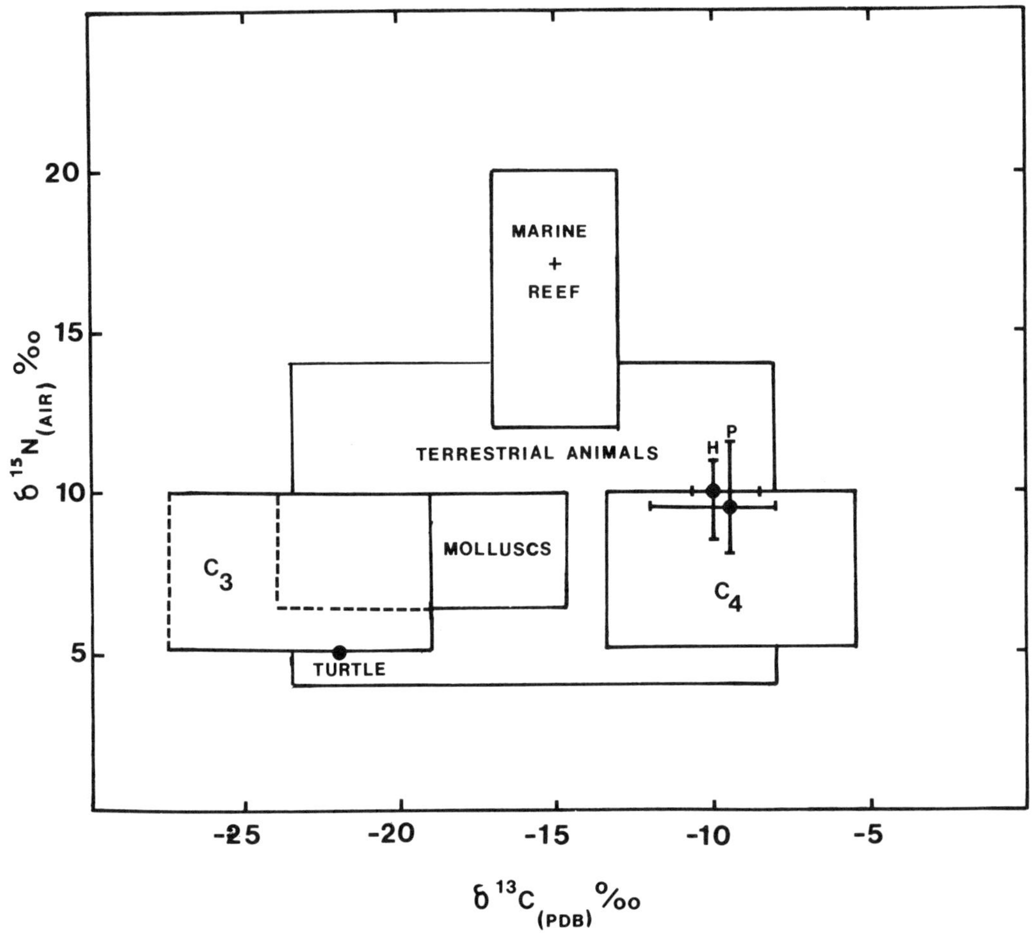

Fig. 1. Model of isotopic range of food resources available at Lamanai versus consumption (P, Postclassic; H, Historic). Data for model from White et al. (1991), White and Schwarcz (1989), Schoeninger et al. (1983), Medaglia et al. (1989), and Reed (1992).

TABLE 2. Comparison of Dietary Indicators Between Time Periods

	Postclassic			Historic		
	n	Mean	SD	n	Mean	SD
$\delta^{13}C$	25	9.3	0.8	11	9.9	0.9
$\delta^{15}N$	24	9.5	0.9	9	9.7	0.6
Caries	50	22.8	40.4	42	20.1	25.7
Calculus	50	1.2	0.7	42	1.2	0.6
Wilson bands	8	0.9	0.6	15	2.4	1.5 *

Isotopic data from White and Schwarcz (1989); dental data from White (1986) and Wright (1990).
*Significant at $p \leq 0.05$.

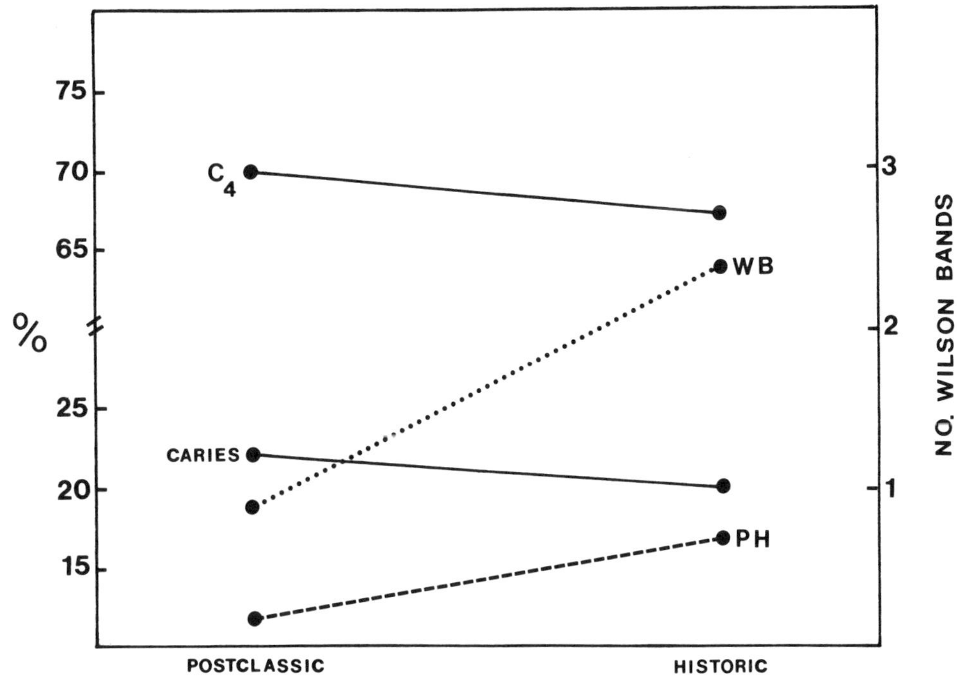

Fig. 2. Comparison of dietary indicators (percent C_4 foods and mean percent caries) with physiological stress indicators (number of Wilson bands [WB] and frequency of porotic hyperostosis [PH]) for Postclassic and Historic periods.

each tooth using the scale of Baer et al. (1961), used for Maya elsewhere (Evans, 1973), and a mean score was determined for each dentition. Although both measures are controlled for age (i.e., only adults over 20), sources of potential error include premortem and postmortem tooth loss, calculus preservation, and preservation factors affecting teeth that are differentially affected by these pathological indicators.

The dependency on maize suggested by $\delta^{13}C$ values is further supported by high frequencies of dental caries and substantial amounts of calculus buildup for both time periods (Table 2). Both indicators remain stable between the Postclassic and Historic periods (t tests, $p > 0.05$).

PHYSIOLOGICAL STRESS

Porotic Hyperostosis

The anemia-related lesions—cribra orbitalia and porotic hyperostosis—have been identified in many Maya populations from the Classic to Postclassic periods (Hooton, 1940; Saul, 1972). These lesions likely represent anemia episodes during childhood (Stuart-Macadam, 1985).

The iron store balance is a result of complex interactions among intake, absorption, and loss; therefore, its etiology in skeletal populations has been a subject of considerable controversy (Carlson et al., 1974; Weinberg, 1974; Stuart-Macadam, 1985; Palkovich, 1987) The physiological, demographic, ecological, cultural, and nutritional conditions affecting the frequency of porotic hyperostosis will vary by region and site.

Regardless of time period, at Lamanai we find a relatively large maize-dependent population living on a lagoon in a tropical environment. Loss of iron through perspiration (WHO, 1968), possible helminthic or other parasitic infection, and heavy maize consumption would put at least some segments of the population at risk of iron deficiency. The $\delta^{15}N$ values do not indicate significant consumption of fish, which can protect against anemia (Layrisse et al., 1969) if they do not contain helminthic parasites.

Physiological differences could account for the patterns of iron-deficiency skeletal lesions in the population structure: there is a greater representation of lesions in groups at greater risk for anemia in the Historic period, namely juveniles. Further demographic work will be needed to confirm these observations, which have been confounded by poor preservation. For example, from 12% to 15% of skeletons were not well enough preserved to identify age and sex, and to determine lesion presence.

There is a highly significant association between time period and the presence of lesions (χ^2, $p < 0.005$), with a more frequent occurrence of anemia in Historic period skeletons (Table 3, Fig. 2). The $\delta^{13}C$ and $\delta^{15}N$ values of skeletons exhibiting lesions do not differ significantly from those without lesions (t tests, $p > 0.05$). This finding, combined with the absence of any observable shift in diet over time, precludes a food-related explanation. Although the sample is too small to be tested statistically by sex or age, it appears that females are no more likely to be affected than males. In addition, both adult and juvenile skeletons show higher lesion frequencies during the Historic period, and they are equally affected.

TABLE 3. Frequency of Porotic Hyperostosis by Time Period, Age, and Sex

	Postclassic		Historic	
	No. of cases	%	No. of cases	%
Active lesions	5	9	17	17
Normal	48	91	83	83
Sex				
Male	0	0	7	17
Female	2	6	6	11
Unknown	2	—	—	—
Age				
Adult	4	5	13	14
Juvenile	1	3	7	13

Enamel Defects

Physiological stress can affect the production of enamel matrix by ameloblasts. Like hypoplasias, Wilson bands are enamel defects that record disturbances to ameloblast physiology, and they are visible as anomalous bands in longitudinal cross sections cut from teeth (Rose 1977, 1979; Condon, 1981: Rose et al., 1985; Marks, 1988; Wright, 1990). The bands are visible throughout the depth of the enamel, that is, at all angles of section to the prism, a feature that differentiates them from nonpathological striae of Retzius (Wright, 1990). Wilson bands are caused by severe acute stress rather than the low-grade chronic stress that produces enamel hypoplasias (Condon, 1981; Rose et al., 1985); accordingly, Wilson bands are likely to be more diagnostic of short-term morbidity than enamel hypoplasias (Wright, 1990).

Because examination of enamel microstructure requires destructive sectioning of teeth, only a subset of burials was selected (Postclassic, 4 males, 4 females; Historic, 7 males, 8 females). These teeth did not exhibit significant attrition on the canines, but were otherwise chosen arbitrarily.

Using scanning electron microscopy, a total of 43 Wilson bands were identified in the 23 canines examined. The sexes were pooled because no differences were detected in the mean number of Wilson bands between males and females in either sample (t tests, $p > 0.05$). Most canines exhibit at least one defect, but the numbers of Wilson bands are not evenly distributed across the samples. The defects were almost three times more common in Historic teeth than in Postclassic teeth (Historic mean = 2.4, Postclassic mean = 0.88, a difference that is highly significant [t test, $p < 0.01$]). No difference was detected between the samples in the developmental age of Wilson bands (Kolmogorov-Smirnov test, $p > 0.05$).

DISCUSSION

In contrast with situations where the Spanish presence resulted in intensified maize production (e.g., southern Atlantic coastal United States; Schoeninger et al., 1990), Lamanai did not experience cultural change sufficient to alter dietary, and presumably agricultural, habits during the early Historic period. Both chemical and dental data indicate strong maize dependency with probable supplementation from land animals, such as deer and peccary, and a few aquatic resources, such as turtle and fish. Consumption patterns indicated by $\delta^{15}N$ and $\delta^{13}C$ do not change, strongly suggesting dietary stability throughout the initial colonial era. The chemical data are supported by measures of dental caries and calculus, which also remain remarkably stable. The food regime of the ancient Maya does not appear to be much different from what is consumed by modern indigenous populations (Béhar, 1968; Benedict and Steggerda, 1936).

The Historic period skeletons from Lamanai, however, show higher lesion frequencies attributable to heavier physiological stress than their Postclassic ancestors. Given the absence of measurable dietary shift, we must look elsewhere for the cause of lesion frequency changes. We suspect changing epidemiologic conditions resulting from the Spanish contact with Lamanai and, perhaps more important, from trade interaction with the greater Spanish presence to the north in Yucatan.

Both porotic hyperostosis and Wilson bands were certainly present in the Lamanai Maya before conquest, but their rise in Historic times indicates one of the following: a greater experience with infection, most likely occurring in childhood; or a greater likelihood of survival from early illness and, hence, the common appearance in the mortality sample of skeletal lesions that take some time to form. The nature of the physiological stress is, however, in question. Although the presence of anemia implies a chronic condition, the patterning of Wilson bands indicates severe and acute episodic stress. Hypoplasia frequencies do not differ between time periods, or by sex and developmental age (White, 1986; Wright, 1990). We interpret this pattern as evidence for a common weaning age (2 to 6 years) and a time when children were particularly at risk for infection. It is possible that the conflicting lengths of stress periods for these indicators are reflecting different kinds of health conditions, both of which occurred in childhood, namely infectious and parasitic diseases.

Although the *Maya Katun Prophecies of Chilam Balam* indicate that the Maya experienced a series of famines and epidemics (e.g., pneumonia) affecting demography in the Yucatan beginning around 1400 (Roys, 1954), epidemics introduced by the Spanish after 1544 are known to have devastated the colonial Maya. These include measles, influenza, yellow fever, possibly typhus, and, most significantly, smallpox. Next to skin diseases (which might reflect illnesses such as measles and smallpox), intestinal complaints comprised the largest category of ailments treated in Maya ethnohistorical medical texts (White, 1985). Epidemiological conditions are illustrated by a Maya lament for earlier days:

> There was then no sickness; they had then no aching bones; they had then no burning chest; they had then no abdominal pains; they had then no consumption; they had then no headache. At that time the course of humanity was orderly. The foreigners made it otherwise when they arrived here. (Roys, 1933:83)

The Yucatan, however, was more intensively occupied by "foreigners" than the more forbidding environment of inland Belize

(Landa, 1566). Mortality from infection there was high (Farriss, 1984), as it was in the neighboring northern Maya lowlands (Gerhard, 1979). The infectious diseases described above could easily have produced the severe, acute stress signaled by Wilson bands. It is important to note, however, that the appearance of Wilson bands denotes recovery from stress. We are looking at the survivors.

The possibility of increased parasitic infection cannot be ruled out as a causative factor for porotic hyperostosis, although infectious diseases could also have induced an iron-withholding response, or interfered with its absorption and temporarily contributed to anemia. Helminthic parasites were present in the Americas in pre-Columbian times (Horne, 1985; Podo-Ledezma, 1985), and they might have been endemic to the Maya lowlands. The parasitic "load" might have increased in Colonial times, favored by ecological conditions at the site. Several serious parasitic infections causing dysentery were introduced by the Spanish and their African slaves (Manter, 1967; Schmidt and Roberts, 1981), but new pathological conditions with the greatest potential contribution to anemia were malaria and hookworm (Bruce-Chwatt, 1965; Dunn, 1965; Friedlander, 1977). Parasitic infections in lowland Central America are today a major health concern, and in some modern Maya villages the infection rate reaches 100% (Scrimshaw and Tejeda, 1970). The rise in anemia seen in the Historic period could reflect increased parasitic infection, particularly of children. Children living on iron-poor maize diets are susceptible to anemia from parasites that tip the balance between high physiological demand and iron availability; nutritionally marginal intake is aggravated by excessive loss (WHO, 1968; Jenkins, 1981). At Lamanai, children were consuming maize in the same quantities as adults and demonstrated no differences in protein sources (White and Schwarcz, 1989). Their diets were marginal, at best, for both their protein and iron requirements. The presence of both hematologically related skeletal lesions and enamel disturbances suggest many illnesses experienced in early childhood.

CONCLUSION

Dental and chemical measures of diet for the Postclassic and Historic periods indicate a heavy dependence on maize with some supplementation from land animals and, possibly, aquatic resources. The absence of significant dietary differences between the two periods suggests that the shift in health status is a consequence of changing epidemiological conditions in the colonial era. In spite of the remarkable stability indicated by both dietary and archaeological data at Lamanai, the inhabitants of this site did not remain unscathed by their interactions with the Spanish. Physiological stress, as indicated by porotic hyperostosis and enamel defects, appears to have been common in the early Historic period. It clearly affected children, probably around the time of weaning, when they were most vulnerable, and it appears to have been both severe and acute as well as chronic in nature. Although protected somewhat by limited intermittent contact at the local level, we suggest that Lamanai saw the introduction of a host of diseases, probably both infectious and parasitic, primarily through their trade with Spanish and Maya groups in the more heavily occupied Yucatan.

ACKNOWLEDGMENTS

We thank Hermann Helmuth, Department of Anthropology, Larry Justus and James Nighswander, Department of Biology, Trent University, for their guidance in osteological and microscopic analyses. The isotopic analysis was done with the support and supervision of Henry Schwarcz, and the technical assistance of Martin Knyf, Department of Geology, McMaster University. Further recognition is given to J. Buikstra, P. Healy, and R. Stairs. Excavations at Lamanai were supported by the Social Sciences and Humanities Research Council of Canada, the Royal Ontario Museum, and private donors. This work was funded by the departments of Anthropology and Biology, Trent University, and a graduate fellowship from the Natural Sciences and Engineering Research Council to C. White.

REFERENCES CITED

Baer PN, Stephen RM, and White CL (1961) Studies on experimental calculus formation in the rat I: Effects of age, sex, strain, high carbohydrate, high protein diet. Journal of Periodontology 32:190–196.

Béhar M (1968) Food and nutrition of the Maya before the conquest and at the present time. In Biomedical Challenges Presented to American Indians. Scientific Publication 165. Washington, DC: Pan American Health Organization, pp. 114–119.

Benedict FG, and Steggerda M (1936) The food of the present day Maya Indians of Yucatan. Carnegie Institution Contributions to American Archaeology 18, vol. 3.

Bruce-Chwatt LJ (1965) Paleogenesis and paleoepidemiology of primate malaria. Bulletin of the World Health Organization 32:363–387.

Carlson DS, Armelagos GJ, and van Gerven DP (1974) Factors influencing the etiology of cribra orbitalia in prehistoric Nubia. Journal of Human Evolution 3:405–410.

Chase DZ, and Chase AF (1988) A Postclassic Perspective: Excavations at the Maya Site of Santa Rita Corozal, Belize. Pre-Columbian Art Research Institute Monograph 4.

Chase AF, and Rice PM (eds.) (1985) The Lowland Maya Postclassic. Austin: University of Texas Press.

Condon KW (1981) Correspondence of Developmental Enamel Defects Between the Mandibular Canine and First Premolar. M.A. thesis, Department of Anthropology, University of Arkansas, Fayetteville.

Crosby AW Jr. (1967) Conquistador y pestilencia: The first New World pandemic and the fall of the great Indian empires. Hispanic American Historical Review 47:321–337.

Crosby AW Jr. (1972) The Columbian Exchange: Biological and Cultural Consequences of 1492. Westport: Greenwood Press.

Dunn FL (1965) On the antiquity of malaria in the western hemisphere. Human Biology 37:385–393.

Emery K (1986) Variation in a Tropical Gastropod Population; Implications for Ancient Maya Subsistence. Senior Honours thesis, Department of Anthropology, Trent University, Peterborough.

Evans DT (1973) A preliminary evaluation of tooth tartar among the preconquest Maya at the Tayasal area, El Peten. American Antiquity 50:102–116.

Farriss NM (1984) Maya Society Under Colonial Rule. Princeton, New Jersey: Princeton University Press.

Friedlander J (1977) Malaria and demography in the lowlands of Mexico: An ethnohistorical approach. In D Landy (ed.): Culture, Disease and Healing. New York: Macmillan, pp. 113–119.

Gerhard P (1979) The Southeast Frontier of New Spain. Princeton: Princeton University Press.

Graham E (1991) Archaeological insights into Colonial Period Maya life at Tipu, Belize. In DH Thomas (ed.): Columbian Consequences, vol. 3, The Spanish Borderlands in Pan-American Perspective. Washington, DC: Smithsonian Institution Press, pp. 319–335.

Graham E, Jones GD, and Kautz RR (1985) Archaeology and ethnohistory on a Spanish colonial frontier: An interim report on the Macal-Tipu Project in Western Belize. In AF Chase and PR Rice (eds.): The Lowland Maya Postclassic. Austin: University of Texas Press, pp. 206–214.

Graham E, and Pendergast DM (1989) Excavations at the Marco Gonzalez site, Ambergris Cay, Belize, 1986. Journal of Field Archaeology 16:1–16.

Graham E, Pendergast DM, and Jones GD (1989) On the fringes on conquest: Maya-Spanish contact in colonial Belize. Science 246:1254–1259.

Harrison PD, and Turner BL (1978) Pre-Hispanic Maya Agriculture. Albuquerque: University of New Mexico Press.

Hooton EA (1940) Skeletons from the Cenote of Sacrifice at Chichen Itza. In CL Hay, RL Linton, SK Lothrop, HL Shapiro, and CG Vaillant (eds.): The Maya and Their Neighbors. New York: Dover, pp. 270–280.

Horne PD (1985) A review of the evidence of human endoparasitism in the pre-Columbian New World through the study of coprolites. Journal of Archaeological Science 12:169–178.

Jenkins CL (1981) Patterns of growth and malnutrition among preschoolers in Belize. American Journal of Physical Anthropology 56:169–178.

Jones GD (1989) Maya Resistance to Spanish Rule: Time and History on a Colonial Frontier. Albuquerque: University of New Mexico Press.

Landa, D de (1566) Relación de las Cosas de Yucatan, trans. W Gates 1978. Yucatan, Before and After Conquest. New York: Dover.

Larsen CS (1987) Bioarchaeological interpretations of subsistence economy and behavior from human skeletal remains. In MB Schiffer (ed.): Advances in Archaeological Method and Theory, vol. 10. San Diego: Academic Press, pp. 339–445.

Layrisse MJ, Cook JD, Martinea C, Roch M, Kuhn IN, Walker RB, and Finch CA (1969) Food iron absorption: A comparison of vegetable and animal foods. Blood *33*:430–433.

López de Cogolludo D (1971) Los Tres Siglos de la Dominación Española, o sea Historia de esta Provincia. Facsimile of 2nd ed., Campeche-Merida 1842–1845: Graz: Akademische Druck und Verlagsanstalt.

Lovell WG (1985) Conquest and Survival in Colonial Guatemala: A Historical Geography of the Cuchumtan Highlands, 1500–1821. Montreal: McGill-Queen's University Press.

Manter HW (1967) Some aspects of the geographical distribution of parasites. Journal of Parasitology *53*:3–9.

Marks MK (1988) Human Dental Enamel Prism Variation Within Wilson Bands and Hypoplasias: A Scanning Electron Microscope Approach. M.A. thesis, Department of Anthropology, University of Arkansas, Fayetteville.

Medaglia C, Little E, and Schoeninger M (1989) Late Woodland diet on Nantucket Island (Massachusetts): A study using stable isotope ratios. Paper presented, Society for American Archaeology, Atlanta.

Palkovich AM (1987) Endemic disease patterns in paleopathology and porotic hyperostosis. American Journal of Physical Anthropology *74*:527–537.

Pendergast DM (1981) Lamanai, Belize: Summary of excavation results, 1974–1980. Journal of Field Archaeology *8*:29–53.

Pendergast DM (1986a). Under Spanish rule: The final chapter in Lamanai's Maya history. Belcast Journal of Belizean Affairs *3*(1–2):1–7.

Pendergast DM (1986b). Stability through change: Lamanai, Belize, from the ninth to seventeenth century. In JA Sabloff and EW Andrews V (eds.): Late Lowland Maya Civilization: Classic to Postclassic. Albuquerque: University of New Mexico Press, pp. 223–249.

Pendergast DM (1991a) Up from the dust: The Central Lowlands Postclassic as seen from Lamanai and Marco Gonzalez, Belize. In F Clancy and PD Harrison (eds.): Vision and Revision in Maya Studies. Albuquerque: University of New Mexico Press, pp. 169–177.

Pendergast DM (1991b) The Southern Maya Lowlands contact experience: The view from Lamanai, Belize. In DH Thomas (ed.): Columbian Consequences, vol. 3, The Spanish Borderlands in Pan-American Perspective. Washington, DC: Smithsonian Institution Press, pp. 319–335.

Pendergast DM, and Graham E (1991) La mezcla de arqueología y etnohistoria: El estudia del Período Hispánico en los sitios de Tipu y Lamanai, Belice. Paper presented, Tercera Mesa Redonda de la Sociedad Española de Estudias Mayas, Gerona, Spain.

Podo-Ledezma LF (1985) Enfermedades transmitidas por el agua y el colapso de la civilización Maya Clásica. Mesoamérica *10*:391–410.

Reed DM (1992) Ancient Mayan diet at Copán, Honduras, as determined through the analysis of stable carbon and nitrogen isotopes. Paper presented, Paleonutrition: The Diet and Health of Prehistoric Americans Conference, Center for Archaeological Investigations, Southern Illinois University at Carbondale.

Rose JC (1977) Defective enamel histology of prehistoric teeth from Illinois. American Journal of Physical Anthropology *46*:439–446.

Rose JC (1979) Morphological variations of enamel prisms within abnormal striae of Retzius. Human Biology *51*:139–151.

Rose JC, Condon KW, and Goodman AH (1985) Diet and dentition: Developmental disturbances. In RI Gilbert and JH Mielke (eds.): The Analysis of Prehistoric Diets. Orlando: Academic Press, pp. 281–305.

Roys RL (1933) The Book of Chilam Balam of Chumayal. Carnegie Institution of Washington Publication 438.

Roys RL (1954) The Maya Katun Prophecies of the Books of Chilam Balam. Carnegie Institution of Washington Publication 606.

Saul FP (1972) The Human Skeletal Remains of Altar de Sacrificios. Papers of the Peabody Museum of Archaeology and Ethnology, Harvard University 63(2).

Schmidt GD, and Roberts LS (1981) Foundations of Parasitology. St. Louis: C.V. Mosby.

Schoeninger MJ, DeNiro MJ, and Tauber H (1983) Stable isotope ratios of bone collagen reflect marine and terrestrial components of prehistoric human diet. Science *220*:1381–1383.

Schoeninger MJ, van der Merwe NJ, Moore K, Lee-Thorp J, and Larsen CS (1990) Decrease in diet quality between the prehistoric and contact periods. In CS Larsen (ed.) The Archaeology of Mission Santa Catalina de Guale: 2. Biocultural Interpretations of a Population in Transition. Anthropological Papers of the American Museum of Natural History 68, pp.78–93.

Scrimshaw NS, and Tejeda C (1970) Pathology of living Indians seen in Guatemala. In TD Stewart (ed.): Handbook of Middle American Indians, vol. 9. Austin: University of Texas Press, pp. 203–225.

Stuart-Macadam P (1985) Porotic hyperostosis: Representative of a childhood condition. American Journal of Physical Anthropology *66*:391–398.

Thompson JEH (1966) The Maya central area at the Spanish Conquest and later: A problem of demography. Proceedings of the Royal Anthropological Institute of Great Britain and Ireland, pp. 23–37.

Weinberg ED (1974) Iron and susceptibility to infectious disease. Science *18*:952–956.

White CD (1985) Paleoepidemiology of the Pre-Columbian Maya. Manuscript on file, Department of Anthropology, Trent University, Peterborough.

White CD (1986) Paleodiet and Nutrition of the Ancient Maya at Lamanai, Belize: A Study of Trace Elements, Stable Isotopes, Nutritional and Dental Pathology. M.A. thesis, Department of Anthropology, Trent University, Peterborough.

White CD, Healy PF, and Schwarcz HP (1991) Intensive agriculture, social status and Maya diet at Pacbitun, Belize. Paper presented, Canadian Association of Physical Anthropologists, Hamilton.

White CD, and Schwarcz HP (1989) Ancient Maya diet at Lamanai, Belize: As inferred from isotopic and chemical analysis of human bone. Journal of Archaeological Science *16*:451–474.

World Health Organization (1968) Nutritional Anemias. Geneva.

Wright LE (1990) Stresses of conquest: A study of Wilson bands and enamel hypoplasias in the Maya of Lamanai, Belize. American Journal of Human Biology *2*:25–35.

Chapter 13

The Biological Impact of European Contact in Ecuador

Douglas H. Ubelaker

Department of Anthropology, National Museum of Natural History, Smithsonian Institution, Washington, DC 20560

Without doubt, Columbus's voyage of 1492 opened the door for European expansion into the Americas and forever changed the nature of the populations on both continents. The stream of Europeans pouring into the Americas in subsequent years introduced diseases that were new to the American aboriginal population (Crosby, 1972). The mortality induced by these diseases, along with associated socioeconomic disruption, declines in fertility, and other factors, greatly reduced the size of the aboriginal population until the early twentieth century (Ubelaker, 1988a).

The level of morbidity and mortality of the aboriginal population of the Americas before and after European contact continues to be debated. Those who believe the Americas were vastly populated prior to 1492 and subsequently decimated argue that in precontact times diseases were almost unknown. Precontact populations were thus allowed rapid, unchecked population growth (Dobyns, 1983). On the other hand, numerous studies of New World archaeologically recovered human remains dating prior to European contact document high infant mortality rates and considerable additional evidence of mortality and morbidity, suggesting the pre-1492 New World condition was not disease free (Verano and Ubelaker, 1992).

The Spanish explorers Ojeda, Francisco Pizarro, and Blasco Nunez de Balboa reached Andean South America in the early sixteenth century. Some epidemic disease may have reached Ecuador prior to 1530, but smallpox epidemics are known to have occurred in Quito in 1557 and subsequently throughout the remainder of the sixteenth century (Alchon, 1991; Ubelaker, 1992a). The early historic documents and subsequent modern analysis suggest heavy mortality in the sixteenth and seventeenth centuries, but little is known from historic records about general levels of morbidity during the early historic period and how they compared with those of precontact times. This issue is explored using several osteological indicators of life experience in five temporally distinct groups of skeletons.

SKELETAL SAMPLES

Research conducted since 1973 on archaeological samples from Ecuador in South America offers an opportunity to assess the patterns of change in mortality and morbidity over the past 8,000 years. This work be-

In the Wake of Contact: Biological Responses to Conquest, pages 147–160

gan with the excavation of a large Late Integration period mortuary complex from the coastal Ayalán site (Ubelaker, 1981). Since then, excavation and study of human remains has resulted in the analysis of about 1,474 skeletons from 20 distinct samples from both coastal and highland areas (Fig. 1, Table 1).

The samples represent professionally excavated and well-dated human remains. Collectively, they span the entire temporal range of Ecuadorean populations. Although sampling problems persist, the data are unique in Latin America for completeness.

To elucidate long-term trends and illuminate biological change correlated with the European presence, skeletons from individual sites are grouped into five temporally distinct samples. These "periods" do not represent previously defined stages of cultural development, but rather, groupings that divide existing biological data on a temporal basis. This approach masks intrasample variability, but addresses sampling problems associated with small collections and provides a first approximation of long-term trends.

The earliest sample, early precontact, consists of 192 skeletons from a preceramic/preagricultural coastal site (Stothert, 1985; Ubelaker, 1980a, 1988b). Stothert (1985:613) describes the Vegas population as "unspecialized hunters, fishermen, and gatherers living in a littoral zone who added plant cultivation to their subsistence system," and there is some evidence for bottle gourd (*Lagenaria siceraria*) and primitive maize (*Zea mays*).

Five sites yielding 291 skeletons are combined into the intermediate precontact period (Buys and Domínguez, 1988a,b, 1990; Ubelaker, 1980b, 1983, 1988c–e, 1990; Valdez, 1987; Villalba, 1988). These sites are located in the highlands (Cotocollao and Cumbayá) as well as along the coast (OGSE-MA-172, La Libertad, and La Tolita). They include the extensively looted mortuary complex of La Tolita where many artifacts of precious metals accompany the interments. By this time, agriculture was well established in both the highlands and the coast, a shift toward increased sedentism had occurred, and population densities were higher.

The late precontact period includes Ayalán (Ubelaker, 1981) and Agua Blanca (Ubelaker, 1988f) on the coast and the shaft tombs of La Florida in the highlands. During this period, Ecuadorean populations were living in dense settlements and relied heavily on agriculture. This sample of 518 individuals dates from as early as 500 B.C. to about the time of contact.

The Ayalán site (Ubelaker, 1981) consists of two components, including large urn burials containing mostly secondary (disarticulated) remains and earlier, mostly primary skeletons. Some Ayalán urn burials may postdate European contact, but most of the sample is precontact.

The very deep shaft tombs of La Florida are thought to be the tombs made for the elite ruling class, but they also contain the remains of slaves and sacrifices (Doyon, 1988).

All of the historic samples originate from research conducted during the last several years at two of the historic churches in Quito, including San Francisco and Santo Domingo. The Convento de San Francisco was founded in 1535 by the Belgian Fray Jodoco Ricke with construction completed in the early seventeenth century. San Francisco continues to function today as a Franciscan monastery and Catholic church in Quito. By 1989, excavations that began in 1987 had produced about 201 human skeletons from areas within the convent that range in age from the initial years of European contact to about A.D. 1858 (Teran de Rodriguez, 1988; Ubelaker, 1992a). Of the pre-1990 excavations at San Francisco, that of the Zaguan is at the outside entrance to

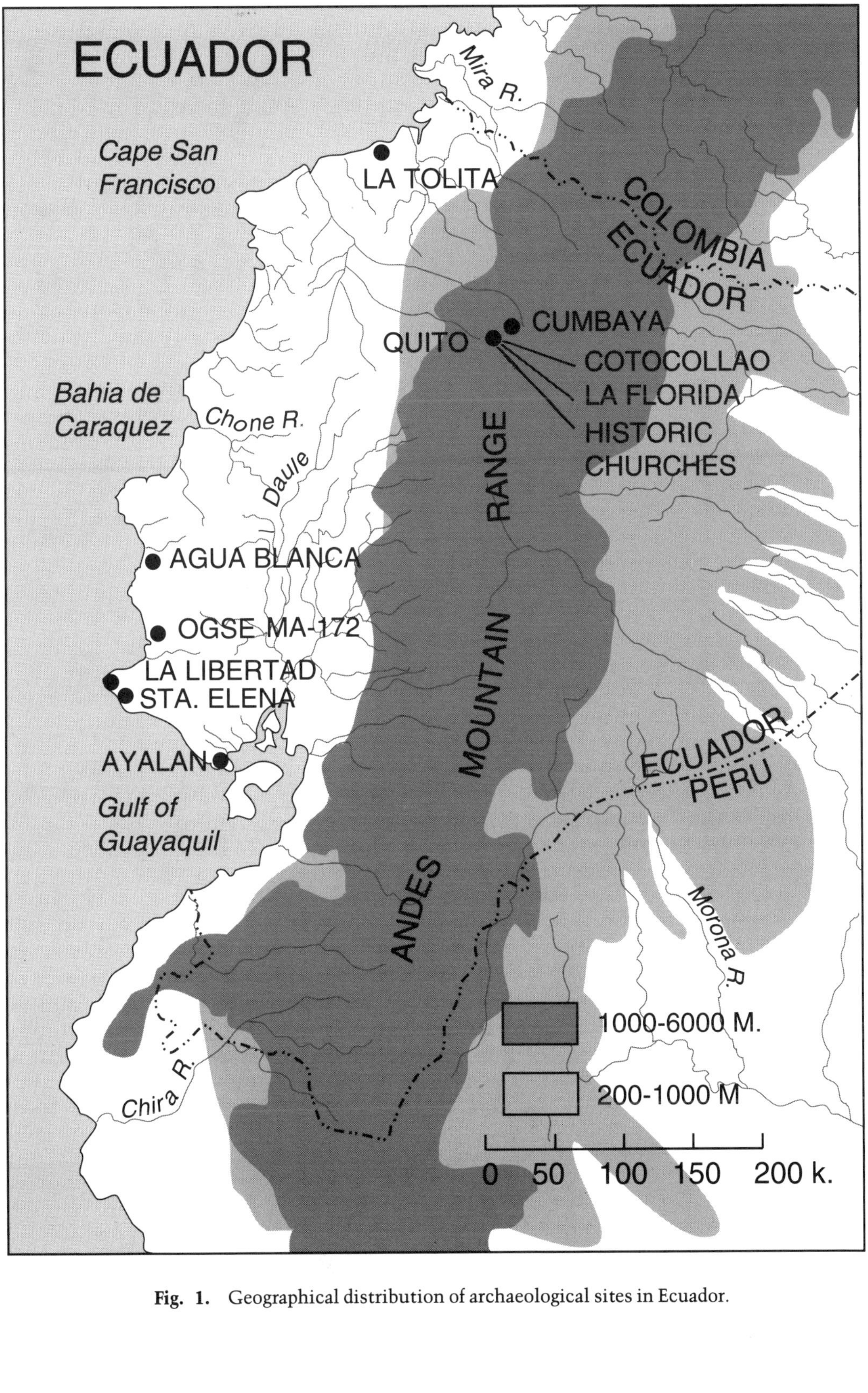

Fig. 1. Geographical distribution of archaeological sites in Ecuador.

TABLE 1. Archaeologically Recovered Samples of Human Skeletal Remains from Ecuador

Period	Sample	Date	Location (Province)	Culture	No. in sample
Early precontact	Sta. Elena	8250 B.P.–6600 B.P.	Guayas	Vegas Complex	192
Intermediate precontact	Cotocollao	1000 B.C.–500 B.C.	Pichincha	Cotocollao	199
	La Libertad (OGSE-46)	900 B.C.–200 B.C.	Guayas	Engory	24
	Cumbaya	400 B.C.–A.D. 100	Pichincha	Cumbaya	20
	OGSE-MA-172	100 B.C.	Guayas	Guangala	30
	La Tolita	A.D. 90–A.D. 190	Esmeralda	Tolita Tardio	18
Late precontact	Ayalán Non-urn	500 B.C.–A.D.1155	Guayas	Milagro	51
	La Florida	A.D. 340	Pichincha	Chaupicruz	76
	Agua Blanca	A.D. 800–A.D. 1500	Manabi	Manteño	7
	Ayalán Urn	A.D. 730–A.D. 1730	Guayas	Milagro	384
Early historic	San Francisco, Zaguan	A.D. 1500–A.D. 1570	Pichincha	Historic	30
	Santo Domingo	A.D. 1500–A.D. 1650	Pichincha	Historic	46
	San Francisco, strata cut, upper level	A.D. 1540–A.D. 1650	Pichincha	Historic	74
	San Francisco, strata cut, lower level	A.D. 1580–A.D. 1700	Pichincha	Historic	46
	San Francisco, Atrium	A.D. 1600–A.D. 1725	Pichincha	Historic	19
Late historic	San Francisco, Church	A.D. 1535–A.D. 1858	Pichincha	Historic	119
	San Francisco, superficial collection, lower level	A.D. 1670–A.D. 1709	Pichincha	Historic	52
	San Francisco, Main Cloister	A.D. 1730–A.D. 1858	Pichincha	Historic	33
	San Francisco, superficial collection, upper level	A.D. 1770–A.D. 1890	Pichincha	Historic	21
	San Francisco, Boxes	A.D. 1850–A.D. 1940	Pichincha	Historic	33

the complex and appears to represent the local Indian population. The other components in this sample (Atrium, Main Cloister, and Church) appear to contain predominantly European remains, although at least some non-European Indian and individuals of African descent are present. The blacks in the series likely represent early slaves or indentured servants who worked for the church.

In 1990, excavations at San Francisco focused on the central ossuary of the church, and they revealed large concentrations of human remains within three chambers located below one of the chapels. Samples were obtained from the superficial layers and strata cuts of each of the concentrations within the two vertically arranged chambers. The contents of 11 individual boxes within a side chamber were also studied.

The ossuaries within San Francisco appear to contain the remains of mostly Europeans, although one individual of apparent African ancestry was discovered within the strata cut sample from the upper chamber. All samples, except the superficial collection from the lower chamber, appear to represent cross sections of the general church population. The exception contained mostly remains of adult males that presumably represent officials of the church who died and were deposited after about A.D. 1770.

The remains that resulted from the excavation in the church of Santo Domingo have not been securely dated. However, they likely originate from the earliest periods of European contact at Quito.

To facilitate analysis of the historic skeletons, the sample has been divided into early and late. The early sample (215 individuals) includes the Santo Domingo material as well as the San Francisco Zaguan and Atrium samples and the strata cuts from the upper and lower levels of the ossuaries. These samples date from as early as A.D. 1500 to probably the late seventeenth century. The late historic sample of 258 individuals probably dates from the last few years of the seventeenth century to the twentieth century. All are from San Francisco, and they include the boxed remains that may date as recently as A.D. 1940.

The historic samples represent a unique and welcome addition to the data available on the biology of Ecuadorean populations. It must be remembered, however, that they are not a representative sample of the entire Ecuadorean population. Because San Francisco church has long been recognized as the most prestigious church in Quito, and probably in the Ecuadorean highlands as well, it likely represents a distinct subsection of Ecuadorean society. In the historic samples, only the Zaguan material of the early historic period represents Indians. The remaining components of the early and late historic period samples represent predominately Europeans with some skeletons of Indian or African descent also present. Because the Zaguan sample is small (30 individuals) and fragmentary, it was grouped with other early historic period samples.

STATURE

Stature was estimated from long bone lengths using standard formulae. Those of Genovés (1967) were utilized for most of the precontact skeletons and the Zaguan historic sample from San Francisco. These formulas were supplemented with those by Trotter (1970) for the historic European material.

Female stature shows little variability through time (Table 2). Mean stature increased 2 cm from early precontact (149 cm) to intermediate precontact (151 cm). It then remained constant until the late historic when the value increased another 3 cm (to 154 cm).

Male stature also did not vary greatly, but it shows a slightly different pattern than that of females. Male stature remained constant throughout the precontact period at 161 cm.

TABLE 2. Temporal Change in Biological Variables

Biological variable	Early precontact	Intermediate precontact	Late precontact	Early historic	Late historic
Female stature (cm)	149	151	151	151	154
Male stature (cm)	161	161	161	155	161
Life expectancy at birth (years)	25	26	23	15	20
Life expectancy at age 5 (years)	28	27	31	23	23
Life expectancy at age 15 (years)	22	23	25	28	23
Ratio of immature individuals to adults	0.57	0.32	0.78	1.29	0.66
Maximum longevity (years)	60	54	68	55	61
Ratio of bones with periosteal lesions to individuals	0.06	0.11	0.06	0.13	0.22
Ratio of bones with periosteal lesions to adults	0.09	0.14	0.12	0.28	0.37
Ratio of bones with trauma to adults	0.09	0.06	0.12	0.12	0.29
Percentage of permanent teeth with carious lesions	3.0	2.7	8.7	5.6	16.6
Percentage of permanent teeth with alveolar abscesses	1.0	1.9	3.1	1.4	4.6
Percentage of permanent teeth lost antemorten	6.0	9.8	11.8	20.0	25.4
Percentage of permanent teeth with hypoplasia	1.0	2.0	4.4	3.0	0.5

It then decreased 6 cm during the early historic period before increasing in the late historic to its previous level.

The precontact lack of variability is somewhat surprising because other indicators of mortality and morbidity do show change. In general, they were short people, and perhaps it can be said that throughout the precontact period sickness and poor nutrition were severe enough that these populations failed to realize their growth potential.

The historic data are enigmatic since they incorporate Europeans as well. It also should be remembered that the historic church samples contain peoples affiliated with the church and do not represent all aspects of Ecuadorean society. The data presented thus combine factors of change in stature brought about by shifts in disease and nutrition along with possible genetic differences in stature. The latter factor may have been minimal since the Spanish also tended to be short. Without doubt, the diet of church officials and high-ranking members of the church was different from common residents of Quito.

DEMOGRAPHY

Although demographic reconstruction from archaeologically recovered human remains offers potentially useful comparative information on life expectancy, mortality, and so forth, interpretation is complicated by numerous potential sources of error. Accurate assessment of demographic change in past populations depends upon adequate sampling, accurate estimation of age-at-death from skeletal analysis, and proper mathematical manipulation of the data. Recent research indicates that mortality statistics derived from skeletal analysis can be influenced by fertility as well as mortality. In the absence of additional information, the relative contributions of fertility and mortality can be difficult to differentiate (Sattenspiel and Harpending, 1983; Buikstra et al., 1986; Johansson and Horowitz, 1986; Milner et al., 1989). Demographic reconstruction also assumes stationary populations when such dynamic variables frequently remain unknown.

The samples from Ecuador summarized in Table 1 are among the largest and best documented in the Western Hemisphere, but still suffer shortcomings. Many of the skeletons were fragmentary or were incomplete secondary deposits. Fertility and the extent of population growth or decline at the time largely remain unknown. Age-at-death was estimated as accurately as possible, but the criteria varied over the 20-year period of analysis. Despite these problems, the data (Table 2) tentatively suggest long-term trends that relate to the various measures of stress and morbidity discussed elsewhere.

Values for life expectancy at birth for these samples increased slightly from early precontact to intermediate precontact, but then dropped sharply during the late precontact, apparently reflecting increases in infant mortality at that time. Presumably, infant mortality increased in response to an increase in disease, related to growing sedentism and population density, as well as a possibly more limited diet.

During the early historic period, life expectancy at birth abruptly declines further to extremely low levels. If these data are correct, they likely reflect the impact of European-introduced epidemics and correlate with the historic record for high mortality.

The late historic period shows a substantial rebound in life expectancy at birth, but with values that remain below those documented for the precontact period.

Life Expectancy at Age 5

Values for life expectancy at age 5 (Table 2) plot mortality in the samples excluding the infants and very young children. The intermediate precontact values decline slightly but then increase during the late precontact. This pattern is opposite to that demonstrated with life expectancy at birth, suggesting that the morbidity problems of the late precontact period focused especially on the infant and young children of the population. The early historic data show sharp decreases without a recovery during the late historic period.

Life Expectancy at Age 15

Plotting life expectancy values at age 15 reveals a slightly different pattern (Table 2). The value actually increases slightly during the precontact periods and also during the early historic period but decreases sharply during the late historic period. These data, coupled with the previous discussed demographic trends, including longevity, argue that the mortality impact during the late precontact and early historic periods predominately focused on the immature individuals in the population. Mortality of adults was significant but remained more stable throughout all periods.

Ratio of Immature Individuals to Adults

This statistic compares the number of individuals less than 15 years of age to those greater than 15 years. The temporal plot of these data (Table 2) clearly demonstrates the disproportionately high impact of disease and other stressors on immature individuals during the late precontact and early historic periods. During the early historic period more deaths occurred among individuals less than 15 years than among those greater than 15 years.

Maximum Longevity

Maximum longevity simply records the oldest likely age for any individual within the sample. For this study, the mean was calculated from the ages of the oldest individuals in each of the five samples. These data are probably the least reliable of any in this study because age-at-death is especially difficult to estimate for the advanced adult years. The problem is compounded because remains from many sites were fragmented and incomplete. Techniques for estimating age-at-death also have varied slightly during the 20 years of study (see References Cited for Ubelaker). For example, ground thin sections were prepared from the femoral midshafts of adults from Ayalán, and ages were estimated through assessment of cortical remodeling. This technique was not used subsequently because most work was completed in Ecuador where slide preparation equipment was not available.

In spite of these caveats, the data suggest only minor fluctuations in maximum longevity through time (Table 2). The mean maximum longevity of the early precontact sample (60 years) is very close to that of the late historic sample (61.2 years). Figures for the intervening three samples range from 54 to 69 years.

SKELETAL AND DENTAL PATHOLOGY

Periosteal Lesions

In this study, periosteal lesions refer to new bone formation on the outer surfaces of long bones. In most cases, areas of abnormal bone proliferation are well remodeled, indicating a skeletal response long before death. Less commonly, the deposits of new bone show little or no evidence of remodeling, indicating that these conditions were active at about the time of death.

The periosteal lesions are assumed to mostly represent areas where infections had caused a response of the periosteum. It is possible that in a minority of cases, trauma may have stimulated new bone formation. In general, however, periosteal lesion frequencies provide an indication of at least some aspects of the population's experience with infectious diseases. These data obviously offer information only on diseases that affect bone and only for individuals who survived long enough for the evidence to be recorded in bone. In short, periosteal lesion frequencies reflect both exposure to specific infectious diseases and survival from an attack by such a disease.

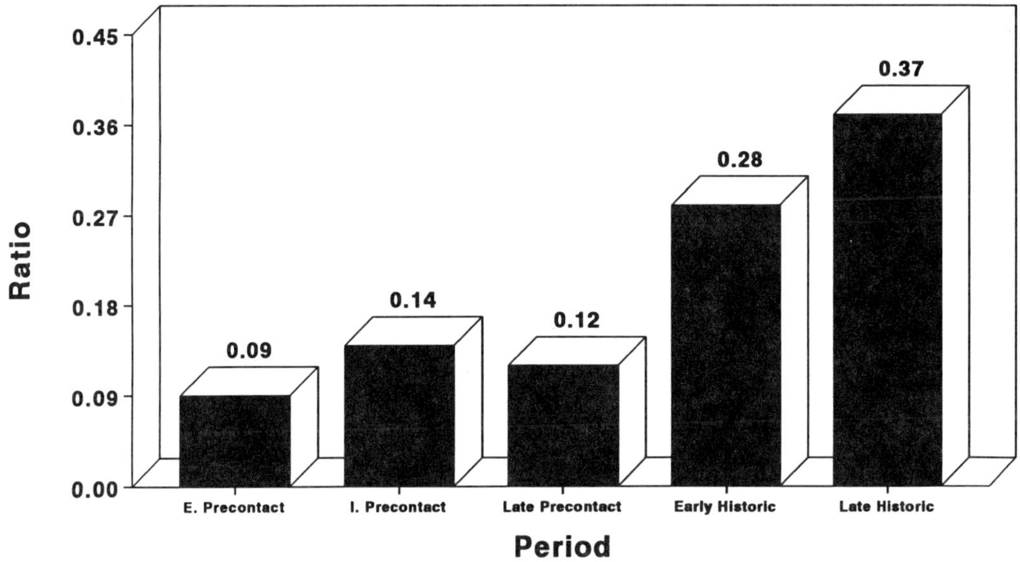

Fig. 2. Temporal change in the ratio of bones with periosteal lesions to adults in the sample.

In the samples from Ecuador, the lesions are found primarily on long bones of adults, whereas we know that subadults suffered greatest mortality from infectious disease. The limited evidence on immature remains may indicate that juveniles tended to die before any evidence of the problem could be left on the skeleton or that such evidence was eliminated by the extensive remodeling involved in the growth of immature bones.

The temporal sequence (Table 2, Fig. 2) reveals low periosteal lesion frequencies in the early precontact period followed by an increase in the intermediate precontact (Table 2). The ratio of bones with lesions to total individuals in the sample is reduced during late precontact, but then it rises dramatically during the early and late historic periods.

The same trend is indicated using only adults, thus eliminating problems posed by unequal representations of adults in the various samples. The decrease during the late precontact, however, is not as dramatic.

Trauma

Evidence for trauma in these samples consists mostly of healed long bone fractures, although some vertebral compression fractures and one example of sharp force trauma were observed. All fractures could clearly be distinguished from postmortem taphonomic change, in part by radiographic analysis.

The temporal trend (Table 2) in trauma frequency reveals relatively low levels of trauma in adults during the early precontact period, a slight decline during the intermediate precontact period, and then an increase during the late precontact period. Trauma frequencies did not change from late precontact to early historic, but then doubled during the late historic period. Presumably, this late historic dramatic increase represents more frequent trauma related to the urban setting.

Carious Lesions

Frequencies of dental caries largely reflect dietary composition, although individual constitution, personal hygiene, and ill health

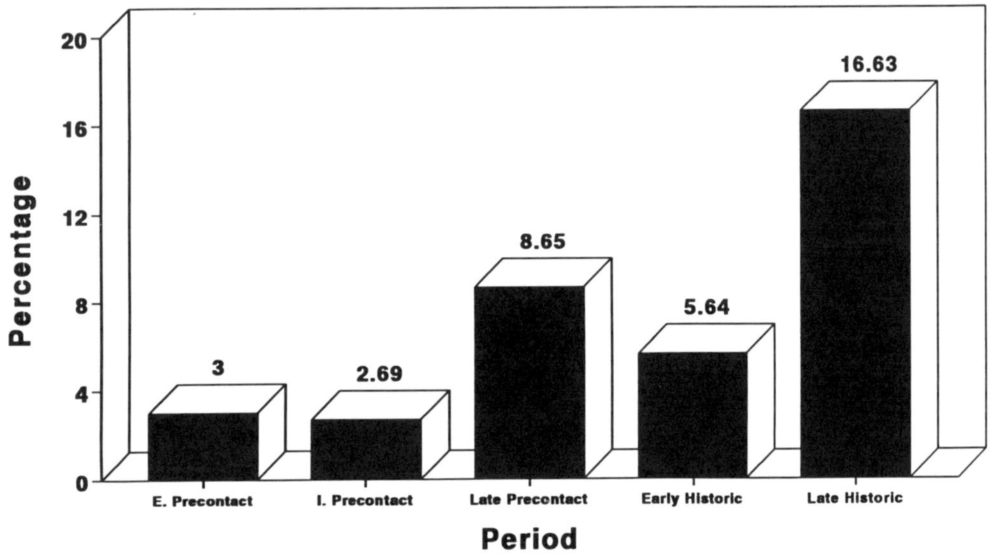

Fig. 3. Temporal change in the frequency of carious lesions in permanent teeth.

can be influencing factors (Newbrun, 1982; Larsen et al., 1991). In this study, carious lesions were defined as areas of tissue collapse on the tooth surface that were at least 1 mm in diameter and appeared to result from caries. This definition excludes all postmortem loss of dental tissue as well as antemortem fractures, chipping, and so forth.

Overall frequencies of carious lesions in fully formed erupted permanent teeth were low in the early and intermediate precontact samples (Table 2, Fig. 3). Frequencies then increased during the late precontact and early historic periods. These data suggest that the impact of a dietary shift toward soft, starchy cariogenic foods did not occur until the late precontact period. Interestingly, the likelihood of dental decay is not uniform among the sites comprising these samples. For example, the highland site of Cotocollao revealed relatively low dental caries frequencies even though it is dated well within the agricultural period when maize, potatoes, and other starchy cariogenic foods formed the focus of Ecuadorean diets.

During the late historic period, dental caries frequency skyrocketed to levels over five times those of the early precontact period. Refined sugar likely was the culprit. Sugar was also available during the early historic period, but it was too costly for the general population. During the late historic period, the sugar prices declined, thus facilitating the regular inclusion of sugar in the diet and consequently the dramatic increase in dental caries prevalence.

Alveolar Abscesses

Alveolar abscesses are osteolytic lesions in the alveolar bone associated with the roots of diseased teeth. Usually, abscesses result in the destruction of supporting bone and loss of infected teeth. Eventually, the bone will be remodelled and the lesion will no longer be apparent.

In all of the Ecuadorean skeletons, the lesions resulted from an exposure of the pulp cavity of the tooth with subsequent infection. Such exposure can result from extreme occlusal attrition, caries, or, rarely, fracture.

In the precontact samples, occlusal attrition produced most abscesses because dental caries were rare and dental attrition was very high. Occlusal attrition often proceeded at a rate faster than secondary dentin could be deposited in response. The first molars were especially vulnerable because they are the teeth that are used in grinding food and they are exposed to masticatory forces longer than later erupting other molars.

During historic times, dental attrition was less rapid, but tooth decay increased. Thus, the principal cause of alveolar abscesses shifted from attrition to dental caries. This pattern is revealed by the percentages of permanent teeth with alveolar abscesses that largely follow those of carious lesions, namely increasing regularly during the precontact periods, declining during the early historic, and then increasing dramatically during the late historic (Table 2).

Teeth Lost Antemortem

Teeth lost antemortem are defined as those permanent teeth that are missing with evidence of alveolar bone remodeling. The remodeling is a necessary prerequisite for identification because in the secondary burial samples many teeth had become detached from the alveolus after death.

The antemortem loss of teeth shows a fourfold increase from early precontact to late historic times. These data reflect the great increases in dental caries and abscesses over time; from the standpoint of tooth loss, these factors outweighed problems posed by extreme attrition during early times.

Enamel Hypoplasia

Enamel hypoplasia has been broadly defined as lines, bands, or pits of decreased enamel thickness observed macroscopically on crown surfaces (Goodman et al., 1984, 1988; Ubelaker, 1992b). The dental literature generally suggests that many levels of physiological stress can result in enamel hypoplasia, thus it is best regarded as a nonspecific indicator of morbidity. In this study, hypoplasia was scored in permanent teeth if a defect involving incomplete enamel formation was visible.

Temporal analysis of the data clearly indicates a dramatic increase in defect frequencies during the precontact periods (Table 2, Fig. 4). Frequencies then decline through the late historic period, which has the lowest frequency of all.

The temporal increase in dental hypoplasia during the precontact periods seems consistent with the various other measures of morbidity. The decline in enamel hypoplasia frequencies during the historic periods is more problematic, especially because other measures of ill health (i.e., periostitis) increase during this period. The apparent decline in these enamel defects likely is correlated with the great increase in dental caries, abscesses, and tooth loss. Enamel hypoplasia may predispose to additional dental disease, especially dental caries. In particular, the irregular development of the tooth surface that is associated with enamel hypoplasia may offer a site for plaque formation and bacterial colonization (Bibby, 1943; Kreshover, 1960; Duray, 1990). Thus, in the Ecuadorean samples frequencies of enamel hypoplasia in the historic permanent teeth could very well have been reduced because the hypoplastic defects became sites for caries-induced destruction, thereby obscuring the original defects and leading to abscesses and tooth loss.

CONCLUSION

Nearly 20 years of research on archaeologically recovered samples of skeletal remains from Ecuador allow a preliminary glimpse of long-term trends in human biology. The recent excavation samples from historic churches in Quito enable patterns of change to be pushed for the first time from the precontact periods into the historic era.

Temporal analysis of the data indicates that most measures of poor health increased during precontact times (Ubelaker, 1984).

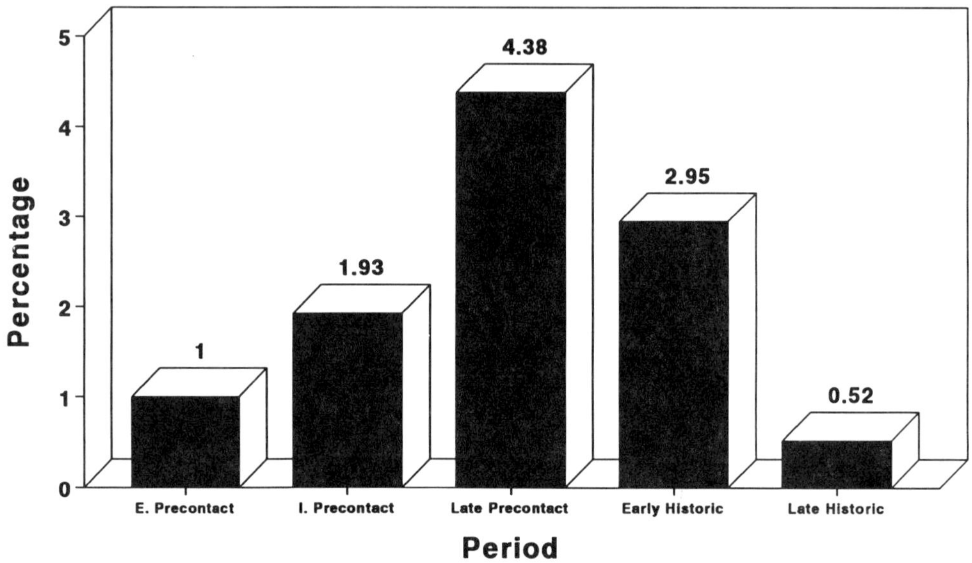

Fig. 4. Temporal change in the percentage of permanent teeth with enamel hypoplasia.

The lowest levels of skeletal indicators of morbidity and, presumably, the healthiest time period for ancient Ecuadoreans was nearly 9,000 years ago during the preceramic, preagricultural occupation on the coast. Skeletal evidence of infectious disease increases during subsequent precontact periods, especially during the late precontact. These increases are coincident with changes in subsistence and, especially, with greater sedentism and population density. In the absence of adequate sanitation, the growing population made less mobile by the change in subsistence economy toward agriculture became increasingly vulnerable to infectious disease. This vulnerability is reflected in the increases during the late precontact period of various measures of juvenile mortality, and also in enamel hypoplasia, a measure of upsets in normal development. Subsistence changes that simultaneously occurred at this time are reflected in an increase in dental disease, especially the frequency of carious lesions and the related problems of alveolar abscesses and antemortem tooth loss.

By the time the Spanish and their new diseases arrived in the sixteenth century, the indigenous populations of Ecuador were already experiencing unprecedented health problems, with the young being especially vulnerable. The demographic impact of the European introduced diseases was rapid and harsh. The high levels of mortality and morbidity experienced during the late precontact periods climbed to dramatic levels and population numbers declined, although adult living stature and maximum longevity changed little during this period. Once again, the mortality among the immature was especially pronounced as the historic periods witnessed declines in life expectancy at birth and at age 5, but slight increases during the early historic for individuals at age 15.

During the historic periods, frequencies of trauma and of periosteal bone lesions, an indicator of exposure to infectious disease, increased dramatically. Dental caries frequencies were at first affected only a little, but then increased dramatically as the people of Quito increased their dietary intake of refined sugar.

This analysis supports the historical evidence that during the early contact periods, life was difficult for the populations of Quito and quite likely for others in Ecuador. New diseases, social upheavals, crowded conditions in the city, and poor sanitation created unprecedented problems for both the Indian and Spanish communities. The data also clearly indicate, however, that the trends toward increasing problems extended back far into the precontact periods. In fact, poor health during the late precontact period likely contributed to the vulnerability of the aboriginal population to the demands of the Spanish and the challenges of new diseases.

ACKNOWLEDGMENTS

The research summarized in this chapter would not have been possible without the cooperation of many archaeologists and administrators associated with the various projects. Especially helpful in this regard have been staffs of the museums of the Banco Central in Guayaquil and Quito, the Instituto Nacional de Patrimonio Cultural del Ecuador, the Cooperacion Tecnica al Desarrolo del Gobierno del Belgica (Belgium), and the Agencia Espanola de Cooperacion Iberoamericano (AECI).

REFERENCES CITED

Alchon, AS (1991) Native Society and Disease in Colonial Ecuador. Cambridge: Cambridge University Press.

Bibby BG (1943) The relationship between microscopic hypoplasia (Mellanby) and dental caries. Journal of Dental Research *22*:218.

Buikstra JE, Konigsberg LW, and Bullington J (1986) Fertility and the development of agriculture in the prehistoric Midwest. American Antiquity 51:528–546.

Buys J, and Domínguez V (1988a) Un cementerio de hace 2000 años: Jardín del Este. In IC Cevallos (ed.): Quito Antes de Benalcazar. Quito: Centro Cultural Artes, pp. 31–50.

Buys J, and Domínguez V (1988b) Hace dos mil años en Cumbayá. Proyecto arqueológico "Jardín del Este", Cumbayá, Provincia de Pichincha, Ecuador, Instituto Nacional de Patrimonio Cultural, Quito.

Buys J, and Domínguez V (1990) El proyecto arqueologico "Jardin del Este," La Preservacion y Promocion del Patrimonio Cultural Del Ecuador. Cooperacion Tecnica Ecuatoriana, Belga 4. Quito: Instituto Nacional de Patrimonio Cultural, pp. 6–21.

Crosby AW (1972) The Columbian Exchange: Biological and Cultural Consequences of 1492. Westport: Greenwood Press.

Dobyns HF (1983) Their Number Become Thinned: Native Population Dynamics in Eastern North America. Knoxville: University of Tennessee Press.

Doyon LG (1988) Tumbas de la nobleza en La Florida. In IC Cevallos (ed.): Quito Antes de Benalcazar. Quito: Centro Cultural Artes, pp. 51–66.

Duray SM (1990) Deciduous enamel defects and caries susceptibility in a prehistoric Ohio population. American Journal of Physical Anthropology *81*:27–34.

Genovés S (1967) Proportionality of the long bones and their relation to stature among Mesoamericans. American Journal of Physical Anthropology *26*:67–77.

Goodman AH, Martin DL, Armelagos GJ, and Clark G (1984) Indications of stress from bone and teeth. In MN Cohen and GJ Armelagos (eds.): Paleopathology at the Origins of Agriculture. Orlando: Academic Press, pp. 13–49.

Goodman AH, Thomas RB, Swedlund AC, and Armelagos GJ (1988) Biocultural perspectives on stress in prehistoric, historical, and contemporary population research. Yearbook of Physical Anthropology *31*:169–202.

Johansson SR, and Horowitz S (1986) Estimating mortality in skeletal populations: Influence of the growth rate on the interpretation of levels and trends during the transition to agriculture. American Journal of Physical Anthropology 71:233–250.

Larsen CS, Shavit R, and Griffin MC (1991) Dental caries evidence for dietary change: An archaeological context. In MA Kelley and CS Larsen (eds.): Advances in Dental Anthropology. New York: Wiley-Liss, pp. 179–202.

Kreshover SJ (1960) Metabolic disturbances in tooth formation. Annals of the New York Academy of Sciences *85*:161–167.

Milner GR, Humpf DA, and Harpending HC (1989) Pattern matching of age-at-death distributions in paleodemographic analysis. American Journal of Physical Anthropology *80*:49–58.

Newbrun E (1982) Sugar and dental caries: A review of human studies. Science *217*:418–423.

Sattenspiel L, and Harpending H (1983) Stable populations and skeletal age. American Antiquity *48*:489–498.

Stothert KE (1985) The preceramic Las Vegas culture of coastal Ecuador. American Antiquity *50*:613–637.

Trotter M (1970) Estimation of stature from intact long limb bones. In TD Stewart (ed): Personal Identification in Mass Disasters. Washington, DC: Smithsonian Institution, pp. 71–83.

Teran de Rodriguez P (1988) Estudio de Investigacion Arquelogica, Convento de San Francisco de Quito, Sitio: OPQSF-2. Report presented to Instituto de Cooperacion Ibero Americana de España, Quito.

Ubelaker DH (1980a) Human skeletal remains from Site OGSE-80, a pre-ceramic site on the Sta. Elena Peninsula, coastal Ecuador. Journal of the Washington Academy of Sciences *70*:3–24.

Ubelaker DH (1980b) Prehistoric human remains from the Cotocollao site, Pichincha Province, Ecuador. Journal of the Washington Academy of Sciences *70*:59–74.

Ubelaker DH (1981) The Ayalán Cemetery: A Late Integration Period Burial Site on the South Coast of Ecuador. Smithsonian Contributions to Anthropology 29.

Ubelaker DH (1983) Human skeletal remains from OGSE-MA 172, an early Guangala cemetery site on the coast of Ecuador. Journal of the Washington Academy of Sciences 73:16–26.

Ubelaker DH (1984) Prehistoric human biology of Ecuador: Possible temporal trends and cultural correlations. In MN Cohen and GJ Armelagos (eds.): Paleopathology at the Origins of Agriculture. Orlando: Academic Press, pp. 491–513.

Ubelaker DH (1988a) North American Indian population size, A.D. 1500 to 1985. American Journal of Physical Anthropology *77*:289–294.

Ubelaker DH (1988b) Restos de esqueletos humanos del sitio OGSE-80. In KE Stothert: Prehistoria Temprana de la Península de Santa Elena, Ecuador: Cultura Las Vegas. Guayaquil: Museos del Banco Central del Ecuador, pp. 105–132.

Ubelaker DH (1988c) Restos humanos prehistoricos del sitio Cotocollao, Provincia del Pichincha, Ecuador. In M. Villalba: Cotocollao: Una Aldea Formativa del Valle de Quito. Miscelania Antropologica Ecuatoriana, Serie Monográfica 2. Quito: Museos del Banco Central del Ecuador, pp. 557–571.

Ubelaker DH (1988d) Human remains from OGSE-46, La Libertad, Guayas Province, Ecuador. Journal of the Washington Academy of Sciences *78*:3–16.

Ubelaker DH (1988e) Prehistoric human biology at La Tolita, Ecuador, a preliminary report. Journal of the Washington Academy of Sciences *78*:23–37.

Ubelaker DH (1988f) A preliminary report of analysis of human remains from Agua Blanca, a prehistoric Late Integration site from coastal Ecuador. Journal of the Washington Academy of Sciences *78*:17–22.

Ubelaker DH (1990) Human skeletal remains from "Jardin del Este," Cumbayá, Pichincha, Ecuador. In La Preservacion y Promocion del Patrimonio Cultural del Ecuador. Cooperación Tecnica Ecuatoriana-Belga 4. Quito: Instituto Nacional de Patrimonio Cultural, pp. 22–52.

Ubelaker DH (1992a) Human Skeletal Biology of Convento de San Francisco de Quito. Report on file, Instituto Nacional de Patrimonio Cultural del Ecuador (INPC) and the Agencia Española de Cooperacion Iberoamerican (AECI), Quito, Ecuador, in press.

Ubelaker DH (1992b) Enamel hypoplasia in ancient Ecuador. In AH Goodman and LL Capasso (eds.): Recent Contributions to the Study of Enamel Developmental Defects. Journal of Paleopathology, Monographic Publications *2*:207–217.

Valdez F (1987) Proyecto Arqueologico "La Tolita" (1983–1986). Quito: Museos del Banco Central del Ecuador.

Verano JW, and Ubelaker DH (eds.) (1992) Disease and Demography in the Americas. Washington, DC: Smithsonian Institution Press.

Villalba M (1988) Cotocollao: Un aldea formativa del valle de Quito. Miscelanea Antropologica Ecuatoriana Serie Monografíca 2. Quito: Museos del Banco Central del Ecuador.

Chapter 14

Biological Effects of European Contact on Easter Island

Douglas W. Owsley, George W. Gill, and Stephen D. Ousley

Department of Anthropology, National Museum of Natural History, Smithsonian Institution, Washington, DC 20560 (D.W.O.); Department of Anthropology, University of Wyoming, Laramie, WY 82071 (G.W.G.); and Department of Anthropology, University of Tennessee, Knoxville, TN 37996 (S.D.O.)

European contact with Easter Island was comparatively recent; that is, in the early eighteenth century. Initial visits tended to be brief, but their frequency increased during the nineteenth century. Ships' logs and other historical accounts provide information about these early contacts and offer some insight into the effects of European contact on the islanders.

Compared to many Polynesian islands, the ethnohistorical documentation of Easter Island is fairly detailed, although it is based on only a few short visits and ethnographic information collected this century from a remnant population. Archaeological evidence has played an essential role in unraveling the human history of this island because of the drastic reduction in population, social disorganization, disruption, and loss of much of the cultural/historical memory that occurred during the postcontact period (Mulloy, 1988). Osteological evidence—bone pathology and osteometric data—is another source of information about the experience, interaction, and effects of contact on this native population.

Human skeletons from recent archaeological investigations on Easter Island date to the late prehistoric and early historic periods. This skeletal series provides an opportunity to determine whether the osteological data support, expand, or contradict the historical record, especially in regard to the transmission of disease, injury resulting from conflict, and genetic admixture. The objectives of this study are the following: 1) to test predictions based on the historical record; 2) to investigate the sociocultural, biological, and health effects of contact with Europeans; and 3) to discover new information that supplements or clarifies the historical record.

PRECONTACT PERIOD

The Pacific island of Rapa Nui (Easter Island), the easternmost outpost of Polynesia, is one of the most remote of the inhabited places on earth. Its nearest neighbor to the west is Pitcairn, 1,819 km distant, and 3,703 km to the east is the coast of southern Peru and northern Chile (Kirch, 1984).

Rapa Nui is a hilly, high, volcanic island without a coral reef or sheltered harbor. At each corner of this triangular island, which measures 20 km by 13 km (180 km^2), are

In the Wake of Contact: Biological Responses to Conquest, pages 161–177

extinct volcanos. The grass-covered landscape is rugged, and at the time of initial European contact it was essentially deforested by the native inhabitants. The aboriginal population depended primarily on water in subterranean pools in lava tubes and on brackish water from coastal springs emptying directly into the sea. Because of the absence of a barrier reef, fishing is poor compared to most other Polynesian islands. With the exception of chickens and rats, introduced during initial colonization, the terrestrial fauna included no birds or mammals.

The island was settled ca. A.D. 400 by Polynesians, and for the next 1,300 years the population increased and developed its culture in isolation (Kirch, 1984). The islanders are known for their remarkable cultural achievements, especially the stone architecture of their large, rectangular, platform altars (*ahu*) and carved monumental statuary (*moi*).

Native subsistence was based on intensive dry-field cultivation, especially of bananas, sweet potatoes, sugarcane, taro, and yams. The primary sources of protein were domestic fowl, seabirds and their eggs, and fish.

By the sixteenth century, population had increased to about 7,000 to 9,000 people (McCoy, 1976; Kirch, 1984). Disruptive social conditions stemmed from overpopulation, environmental degradation from misuse of land, an impoverished natural environment, and the lack of alternative places to live. Conflicts during the late prehistoric and early historic periods forced dispersal from traditional tribal territories and reduced the population size.

Archaeological evidence for warfare dates from the sixteenth century and includes the appearance of fortifications (e.g., cave entrances modified to allow only single-file passage) and flaked obsidian spearheads, as well as less emphasis on the construction of new statues and platforms (Kirch, 1984). Initially, this warfare was probably small scale and sporadic. Radiocarbon age determinations for the second construction phase of *Ahu Akivi*, which has seven statues of "average" size (i.e., 4 m tall and weighing from 12 to 14 tons), indicate that the images were placed on the platform in the early 1600s, probably before A.D. 1660 (Ayres, n.d.). Chronic, widespread warfare would have prevented the successful completion of a public works project of this magnitude.

Warfare is a prevalent theme in local folklore. Ethnohistorical reference to this era indicates social disintegration and restructuring of the political organization, with secular power appropriated by a ruthless warrior class. Chronic warfare among opposing tribes, especially rivalry between the eastern and western tribes, involved frequent raiding, enslavement of those captured, murder, and cannibalism.The primary weapons were stones, small lances or obsidian-tipped spears, and short, heavy, wooden clubs (Routledge, 1919; Metraux, 1940). During warfare, the *moi* of rival groups were deliberately toppled and broken; the last standing *moi* was toppled ca. 1840 (Englert, 1970).

Endemic warfare seems to have been induced by shortages of land and water. Groups with the poorest lands, following the exhaustion of soils, were forced to encroach on the potentially more fertile areas of other groups. The widespread use of fire in raids contributed to the destruction of forests and ground cover that, in turn, exacerbated problems with surface erosion and further reduced the availability of wood used in boats needed for ocean fishing. Periods of social upheaval continued into early postcontact times.

EUROPEAN DISCOVERY, EXPLORATION, AND EXPLOITATION

The first European to discover the island, on Easter Sunday, 1722, was the Dutch navigator Jacob Roggeveen. His ships anchored off the north side of the island for a week during which a landing party of some 150 men went ashore for a day (Routledge, 1919; Metraux, 1940).

The next visits occurred roughly half a century later when ships from Spain (1770), England (1774 and 1795), and France (1786) stopped briefly at the island (Porteous, 1981). From initial contact in 1722 to 1862, 53 visits were recorded (McCall, 1976). Because island resources were limited, there was little to "induce ships that are not in the utmost distress to touch at this island" (James Cook in Porteous, 1981: 11). The greatest resources of the island were the people themselves, whom the Europeans viewed as sources of labor and, in the case of women, sexual satisfaction (McCall, 1976).

Several accounts report the trafficking of women, some participating willingly in exchange for goods (e.g., clothing), with others resisting. Accounts of the visit of the HMS *Blossom* in 1825 indicate that the ship was greeted by 30 or more women sitting "like seals" on a rock in Cook's Bay off Hanga Roa and ". . . singing and accompanying the voice with the arms like sirens alluring one into danger" (Belcher, 1825–1827). La Perouse (1968) wrote indignantly about men forcing young girls, 13 or 14 years old, on the French. The head man of a family often used his female relations as decoys or attention getters, as well as tokens in barter (Van Tilburg, 1992).

Easter Island became known to sailors for the charms of its women. "They were neither reserved or chaste, and for the trifling consideration of a small piece of cloth, some of our sailors obtained the gratification of their desires" (Forster 1777:572). Sexual contact spread venereal disease, and by 1830, ships' captains were cautious of landing on the island because of syphilis (Metraux, 1940). Liaisons, even from the time of the first encounter with the Dutch, also resulted in children who, by island standards, had light skin (McCall, 1981).

The islanders were attracted by European clothing, implements, and other items, and at the time of Roggeveen's contact, their attempt to steal led to "a scuffle . . . over the sanctity of property" (Routledge, 1919:201). The natives picked up stones, and in defense (probably panic or overreaction) a petty officer gave the order to fire on them. Nine or 10 islanders were killed; the rest fled (Routledge, 1919; Metraux, 1940).

Subsequent accounts also report the use of firearms against the islanders. For example, in 1822, the first mate of the whaling ship *Pindos* demonstrated his marksmanship with a rifle by shooting an islander on shore (McCall, 1976).

The experiences of various ship crews on arriving at the island differed, depending on what had occurred during a preceding contact. Encounters resulting in conflict or death (1722, 1774, 1815, 1822, and 1825) were followed by resistance. Of 50 European accounts, fear of attack was mentioned in seven; in five, attacks reportedly occurred (McCall, 1976). La Perouse, for example, experienced some pelting by stones in 1786 when some of his crew attempted to recover a grapnel stolen from the ship. However, it was not until the visits of Captain Adams in 1806 and Captain Windship in 1809 that the islanders repelled visitors by attacking them. Thus, some found the islanders hostile, or were not allowed to land; whereas others, even in the same year, met with no difficulty and maintained peaceful relations with the inhabitants.

An incident in 1805 involving the schooner *Nancy* has been used as the primary explanation for subsequent hostility against Europeans. After a bloody fight with the islanders, the crew kidnapped 12 men and 10 women to help with seal hunting. At the first opportunity, three days out at sea, most of the men jumped overboard to escape and drowned (Metraux, 1940; McCall, 1976).

Whaling vessels also occasionally replaced crew members by kidnapping islanders (Porteous, 1981). During this era of conflict and hostility, only one European was reportedly killed on the island (Hamilton, 1856).

In the interval from December 1862 through March 1863, as many as 1,400 islanders, about 34% of the population, were captured by Peruvian and Spanish slavers for forced labor harvesting guano, primarily on the Chincha Islands located 20 km off the port of Pisco, Peru (McCall, 1976; Boutilier, 1992; Gonnell, 1992). Most, 90%, died from the harsh conditions, unfamiliar environment, and disease. Because of international protest, the trade was officially stopped the following August, and the surviving Polynesians were returned to their homes. However, having contracted smallpox and tuberculosis, only 15 islanders were actually repatriated, bringing the diseases with them (Routledge, 1919, Metraux, 1957; Boutilier, 1992). Smallpox subsequently decimated the island, killing an estimated 1,000 individuals. By 1877, as the result of introduced diseases and of strife between islanders and outsiders, the native population had been further reduced to only 111 people.

Although in effect for less than a year, the Peruvian labor trade brought unparalleled misery and havoc on a number of Polynesian populations. Easter Island, because of its relative proximity to South America, suffered most (McCall, 1976). Leaders and virtually all elders who were repositories of island history, traditions, and culture died, and much of the native culture was lost with them.

OSTEOLOGICAL SAMPLE AND ANALYSIS

The sample numbers nearly 500 complete and partial skeletons, primarily of adults. As a result of secondary burial practices and, in some instances, early collecting preferences, the remains are generally incomplete and represented mainly by crania and long bones. A small collection ($N = 33$) from the Norwegian Expedition (Murrill, 1968) and a majority of the large collection of the Easter Island Museum ($N = 394$) have documented archaeological provenance (Gill, 1986a; Gill and Owsley, 1993). Most of the burials date to the Late Prehistoric (A.D. 1680–1722) and Protohistoric (A.D. 1722–1868) periods (Gill et al., 1983; Gill and Owsley, 1993).

Examination of the collections yielded demographic information, bone inventories, cranial measurements, frequencies of pathological conditions, and anthroposcopic observations for both discrete and continuous traits. Measurements were taken in accordance with procedures defined by Bass (1987), Gill (1971), Gill et al. (1988), and Howells (1973). Anthroposcopic traits were evaluated according to procedures outlined in El-Najjar and McWilliams (1978), Gill (1971, 1986a,b), Houghton (1977), and Jantz (1970). Sex determination for crania used in the metric analysis was based on morphological criteria, including postcranial and pelvic features when available. These assessments were cross-checked through multivariate discriminant analysis. The pathological bone data reported here are limited to traumatic injuries involving crania and periosteal inflammation of the femora, tibiae, and fibulae. In addition, craniometric data are used to detect individual "outliers"; that is, those with morphological characteristics indicative of genetic admixture with Europeans.

BIOLOGICAL RESULTS OF CONTACT

Craniofacial Injuries

Levels of violence can be assessed from frequencies of traumatic injuries (Table 1). Of 2,618 bones examined, 2.5% show either antemortem fractures with evidence of healing or perimortem breakage caused by traumatic injuries. Most injuries were restricted to the cranial vault, particularly the

TABLE 1. Cranial Bone Fractures in Easter Islanders

				Sex						Age					
				Male			Female			15–34			35+		
	N	No. Fx[a]	%	N	No. Fx	%	N	No. Fx	%	N	No. Fx	%	N	No. Fx	%
Frontal	271	31	11.4	141	22	15.6	130	9	6.9	106	14	13.2	165	17	10.3
Left parietal	261	12	4.6	136	10	7.4	125	2	1.6	105	7	6.7	156	5	3.2
Right parietal	269	14	5.2	139	7	5.0	130	7	5.4	107	6	5.6	162	8	4.9
Left temporal	246	2	0.8	134	2	1.5	112	0	0.0	99	1	1.0	147	1	0.7
Right temporal	237	0	0.0	122	0	0.0	115	0	0.0	95	0	0.0	142	0	0.0
Occipital	255	2	0.8	131	1	0.8	124	1	0.8	103	0	0.0	152	2	1.3
Left maxilla	227	1	0.4	120	1	0.8	107	0	0.0	87	0	0.0	140	1	0.7
Right maxilla	228	2	0.9	121	2	1.7	107	0	0.0	88	0	0.0	140	2	1.4
Left zygomatic	212	0	0.0	109	0	0.0	103	0	0.0	85	0	0.0	127	0	0.0
Right zygomatic	215	0	0.0	115	0	0.0	100	0	0.0	83	0	0.0	132	0	0.0
Mandible	197	1	0.5	95	1	1.1	102	0	0.0	67	0	0.0	130	1	0.8
Total	2,618	65	2.5	1,363	46	3.4	1,255	19	1.5	1,025	28	2.7	1,593	37	2.3

[a]Fx, number of fractured bones.

frontal and parietal bones. Slightly more than 11% of frontal bones show evidence of trauma; the frequency for males is more than double that for females. This discrepancy between males and females is also apparent when all cranial bones are considered together.

Most cranial injuries are healed, roughly circular or oval, depression fractures consistent with those caused by a thrown or handheld rock. A few such fractures reflect the shapes of wooden clubs used as weapons by the islanders; others are incisions from blades, occasionally with obsidian fragments still embedded in bone. Such depression fractures and other types of injuries represent interpersonal or intertribal violence among the islanders.

Relatively few cranial injuries were fatal, although there were exceptions. For example, Figure 1 shows an oval, depressed fracture in the frontal bone of a woman (AMNH No. 9809) aged about 55 to 65 years. The defect measures 32 mm by 19 mm and has depressed bone fragments adhering to the margins. A concentric fracture surrounds the central area of penetration, which is missing bone, and multiple chevron-shaped fragments are present along the anterior margin of the depression. Three linear fracture lines radiate outward from the central defect. Endocranially, exfoliated bone and depressed adhering bone fragments are evident along the margin of the lesion. This depressed defect resulted from a single blow. In addition, much of the left parietal, the basilar portion of the occipital, and the temporal bone are missing, probably because of one or more blows to the left side of the head. The fractures are consistent with perimortem trauma and display no evidence of healing.

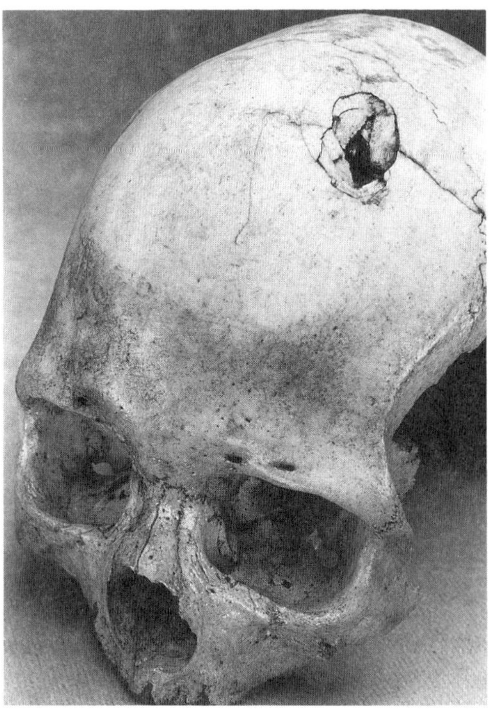

Fig. 1. Fracture of the frontal bone of an adult female (AMNH No. 9809).

One individual sustained a gunshot wound, which reflects a hostile encounter with Europeans. The skull (AMNH No. 9822) of a male aged 25 to 29 years has multiple lead-shot injuries in the forehead (Fig. 2). Three small pellets remain embedded in the frontal bone; two are flattened and measure 6 mm and 7 mm in diameter. A third, located in the superior left half of the frontal, shows less distortion and measures 3.5 mm in diameter. The latter is embedded within a small resorptive crater measuring 5 mm across. Three depressions in the frontal and one in the left parietal mark additional locations where shot struck the cranium. None of the pellets penetrated the inner table of the skull. The man survived, as the wounds are healed. Several accounts mention the use of small shot when firing on islanders (Metraux, 1957; La Perouse, 1968; Forster, 1982; von Saher, 1992).

Transmission of Disease

Historical accounts identify three diseases that were introduced through European contact: smallpox, syphilis, and tuberculo-

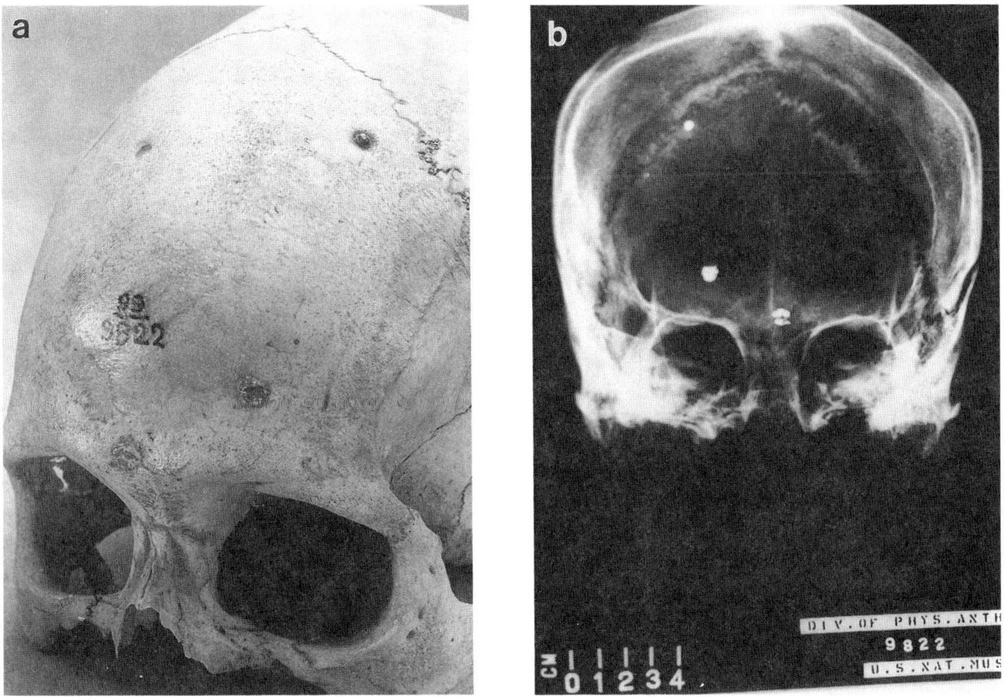

Fig. 2. Embedded pellets from a gunshot wound in the skull (**a**), and X-ray (**b**) (AMNH No. 9822).

sis. Tuberculosis and smallpox reportedly were late introductions brought by returning islanders who survived Peruvian enslavement. Venereal disease preceded these maladies; syphilis was a concern of ships' captains by the early nineteenth century. Pathological changes caused by this disease are evident in the skeletal remains.

Skeletal involvement of syphilis, one of several treponemal infectious diseases, begins in the tertiary stage of the disease, with lesions usually developing between two and 10 years after initial infection. Any bone may be the seat of a syphilitic lesion, but the tibiae, cranial vault, and bones surrounding the nasal cavity are most often affected (Ortner and Putschar, 1981). The most characteristic cranial lesion is caries sicca (Hackett, 1976), a condition of raised smooth nodules surrounding singular or coalesced areas of bone loss. These areas of bone destruction and subsequent healing are said to have a worm-eaten appearance (Mann and Murphy, 1990). Although classic cranial lesions were not identified in the Easter Island crania, one frontal bone has subtle changes reminiscent of stellate scarring.

Periosteal inflammation of the tibiae may result in forward bowing and thickening of the anterior segment of the shaft, sometimes referred to as a saber-shin (Zimmerman and Kelley, 1982). Table 2 lists the frequency of femora, tibiae, and fibulae with pathological thickening of cortical bone. Only the tibiae show remarkably frequent involvement, and male bones present lesions more often than female bones (8% versus 4.6%, respectively).

Causes of periostitis include localized trauma, venous insufficiency, and other

TABLE 2. Periostitis of the Femora, Tibiae, and Fibulae

				Sex						Age					
	Total			Male			Female			15–34			35 +		
	N	P[a]	%	N	P	%	N	P	%	N	P	%	N	P	%
Left femur	187	1	0.5	101	0	0.0	86	1	1.2	35	0	0.0	152	1	0.7
Right femur	183	1	0.5	95	0	0.0	88	1	1.1	37	1	2.7	146	0	0.0
Left tibia	154	11	7.1	85	7	8.2	69	4	5.8	31	2	6.5	123	9	7.3
Right tibia	153	9	5.9	78	6	7.7	75	3	4.0	31	1	3.2	122	8	6.6
Left fibula	103	0	0.0	60	0	0.0	43	0	0.0	26	0	0.0	77	0	0.0
Right fibula	106	2	1.9	65	2	3.1	41	0	0.0	20	0	0.0	86	2	2.3

[a]P, number of bones with periostitis.

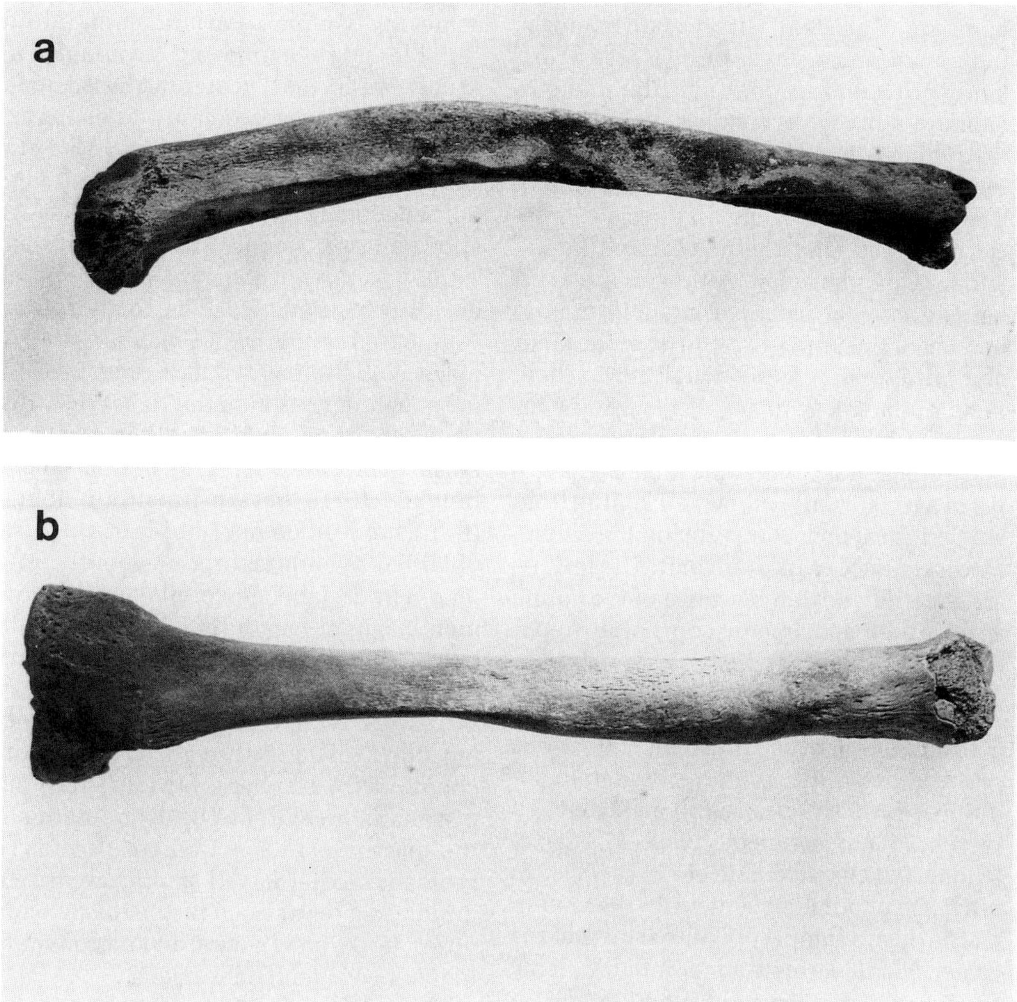

Fig. 3. Right tibiae with bowing (RH0235) (**a**) and thickening of cortical bone (RH0122) (**b**).

causes, so only a subset of the bones identified in Table 2 can be attributed to treponematosis. However, several islanders show pathological changes in the tibiae that suggest treponemal infection, including subperiosteal thickening of the anteromedial and anterolateral diaphyses. This new bone formation often gives the appearance of mild bowing of the anterior diaphysis. These inflammatory conditions are generally well healed.

Figure 3 shows the right tibia of a male 35 to 44 years old (RH0122). The distal half of the diaphysis exhibits new periosteal bone formation on all surfaces. The proximal half of the bone is grossly normal, the shaft is straight, and there is no evidence of cloacae. For the most part, the surface is smooth, but some portions show shallow, parallel striations. Radiographs show involvement of the medullary cavity. The lesion location and overall appearance of the bone are consistent with treponematosis, but nonspecific osteomyelitis is also a possibility. The corresponding fibula shows similar pathological changes.

Another example (Fig. 3) is the right tibia of a female aged 35 to 39 years (RH0235). The middle third of the diaphysis is anteriorly curved, and the medial and lateral surfaces are thickened. Although the main bowing is anterior, there is slight medial bowing of the upper two-thirds of the shaft and mild torsion of the distal third. There is no apparent evidence of cloacae or infection. Both tibiae are similarly affected.

Although not counted in Table 2, the tibiae of a few juveniles show thickening from subperiosteal bone formation on the anteromedial surface of the midshaft. The lesions are generally healed. Because of the young age of these specimens, congenital treponematosis is a possibility.

Genetic Heritage

Examination of the museum collections revealed one skull (1956 Norwegian Expedition, No. 29746) of an adult male that had morphological features atypical of Easter Islanders but characteristic of Caucasians. To provide a population context for the craniometric data, a sample of Caucasians and the Easter Island series were compared using a two-group discriminant function, which provides a means of classifying individual crania according to their degree of similarity to each group. The white sample (80 males, 33 females) represents individuals born during the nineteenth century from the Terry and Todd anatomical collections or other historic samples.

In the two databases, approximately 30 measurements were shared in common after controlling for comparable measurement technique. The measurements were entered into separate male and female stepwise discriminant analyses using SAS PROC STEPDISC (SAS, 1985). This procedure identifies the variables that perform best in achieving sample discrimination, adding variables one at a time until a variable fails to enhance further separation. Based on the results, eight measurements for males and nine for females were used in SAS PROC DISCRIM to develop the discriminant functions. Reducing the number of variables also allowed larger sample sizes for the island series because not all skulls were complete for all measurements. The final analyses were based on 52 island males and 42 females with the required measurements. The measurements used in the linear discriminant functions were as follows: males—biasterionic breadth (ASB), basion-bregma height (BBH), basion-prosthion length (BPL), brow difference (mid-browridge size obtained by subtracting nasion-occipital length from glabella-occipital length), foramen magnum length (FOL), nasal breadth (NLB), bimaxillary subtense (SSS), and maximum cranial breadth (XCB); females—BPL, BBH, NLB, brow difference, biorbital breadth (EKB), orbital height (OBH), simotic chord (WNB), stephanic breadth (STB), and zygomatic breadth (ZYB). Examination of the discriminant function coefficients reveals that Easter Island skulls have greater middle and lower facial prognathism, wider noses, larger brow ridges, and narrower and higher heads than Caucasians.

The resulting functions discriminate between the Easter Island and Caucasian samples very well, especially for females. The Mahalanobis' D^2 between females was 39.1; the value for males was 19.9. Table 3 presents the resubstitution and cross-validation classification summaries for these analyses. Resubstitution uses the total sample to calculate the discriminant function and then classifies each individual. A jackknife or cross-validation analysis sequentially removes each individual and recalculates the function based on the remaining members. The removed specimen is evaluated by the corresponding function. Classification accuracy usually decreases

TABLE 3. Resubstitution and Cross-validation Classifications Using Linear Discriminant Functions (Number and Percentage of Individuals Classified into Each Group)

	Resubstitution				Cross-validation			
	Caucasian		Islander		Caucasian		Islander	
	N	%	N	%	N	%	N	%
Males								
Caucasian	80	100.00	0	0.00	80	100.00	0	0.00
Islander	2	3.85	50	96.15	3	5.77	49	94.23
Females								
Caucasian	33	100.00	0	0.00	33	100.00	0	0.00
Islander	0	0.00	42	100.00	0	0.00	42	100.00

when the specimen being evaluated is excluded from the reference set in this way. All females were correctly classified using resubstitution and cross-validation. With varying degrees of probability, two Easter Island males classified as white using resubstitution, and three using cross-validation.

The white posterior probabilities for these three males were 0.99 (Easter Island Museum No. RH0064), 0.66 (National Museum of Natural History-Santiago, No. 29746), and 0.55 (AMNH No. 9819), which reflect the probability of classification as Caucasian using cross-validation. Posterior probabilities evaluate classifications based on the assumption that the specimen comes from one of the calibration samples (Campbell, 1984). The posterior probabilities for the white male sample show clear separation: all of the values are above 0.96, except for one that is 0.88. The values for females of both samples are even higher (none are less than 0.99), which indicates larger metric differences in craniofacial morphology, less within-group variation, and easier separation by the discriminant function. The discriminant score distributions are illustrated in Figure 4.

The three crania that classified as white were further evaluated with morphological traits useful in distinguishing whites from Polynesians, including six midfacial and interorbital measurements obtained with a simometer (Gill 1984, 1986b; Gill et al., 1988). These measurements are used to construct the maxillofrontal (MF), zygoorbital (ZY), and alpha (AL) indices, which through previous experience have been shown to be useful in distinguishing white from nonwhite crania.

Norwegian Expedition specimen No. 29746 from the site of Hekii consists of a complete cranium of a male, 20–24 years old, without the mandible or associated postcrania (Number 6 in Murrill [1968]) (see Fig. 5). Results of the simometer measurements of midfacial projection are as follows: the first index (MF) scores strongly in the nonwhite sector; the second index (ZY) has an intermediate value; and the last index (AL) falls strongly into the white sector. Overall, the midface falls slightly into the "non-Caucasoid" side, but with an unusual pattern that suggests admixture. Caucasoid traits include large nasal bones with a straight profile, a "steepled" nasal bone form (Brues, 1990), the form and slope of the frontal processes of the maxillae, a marked degree of midline prominence, a parabolic palate, a jagged (Caucasoid) palatine suture form, a strongly curved (quite

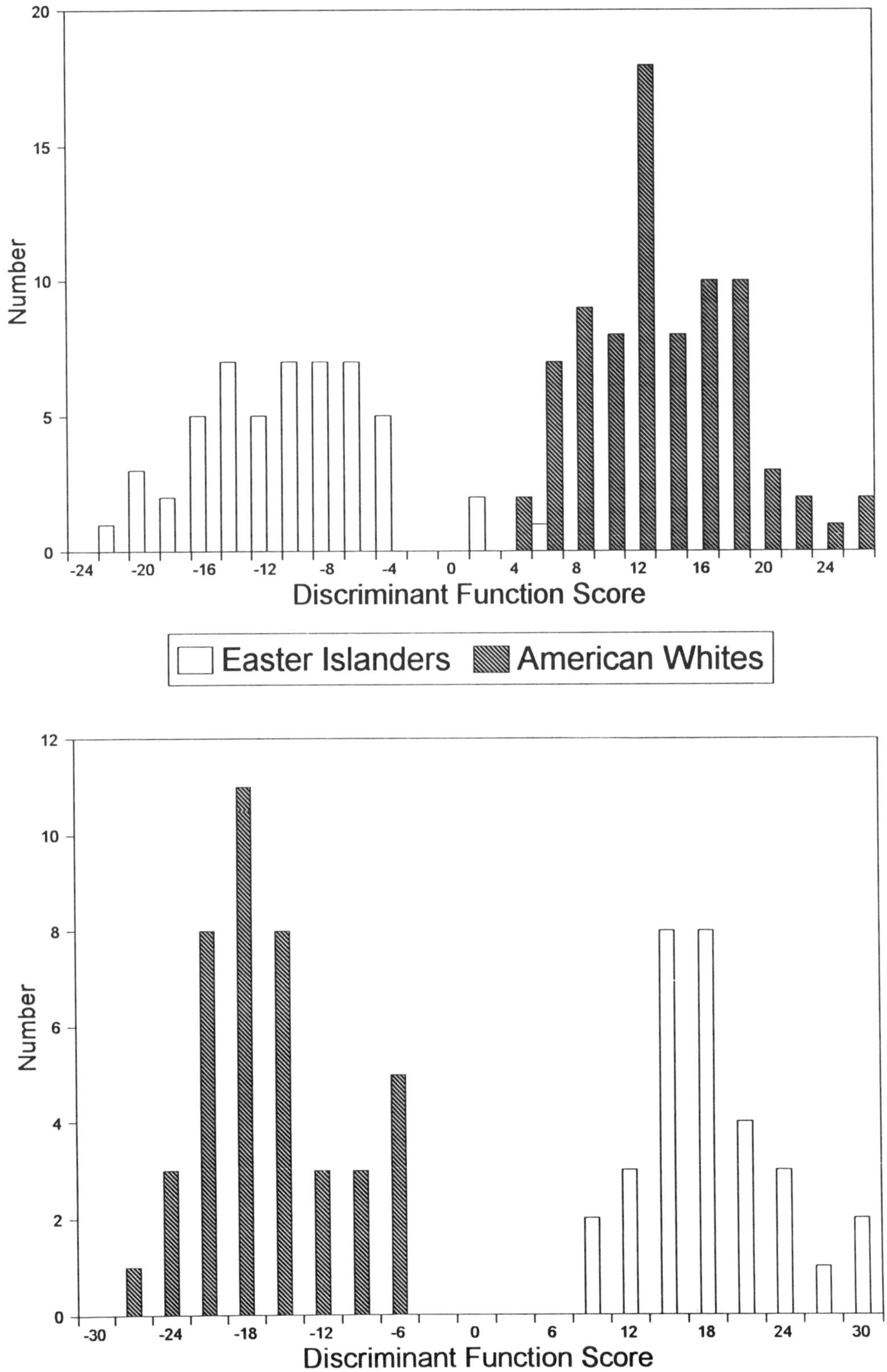
Number
Discriminant Function Score
Easter Islanders
American Whites
Number
Discriminant Function Score

Figure 4.

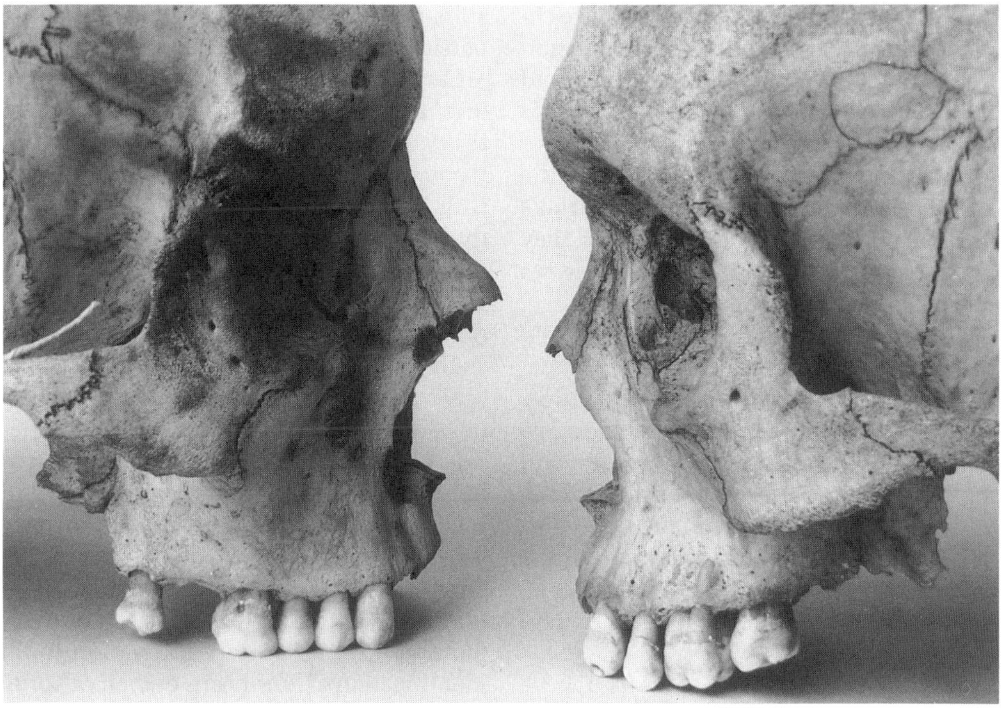

Fig. 5. Lateral view of cranium Norwegian Expedition No. 29746 showing prominence of the nasal bones (left) and a straight facial profile relative to a typical islander (right).

Caucasoid) zygomaticomaxillary suture form, a large nasal spine, a leptorrhine nose (41.7 nasal index), and a shorter, wider cranial base than typical of Easter Island males.

Traits that are east Polynesian and not Caucasoid include a medium-dull nasal sill, the presence of a partial metopic suture line, a visible premaxillary suture, and large numbers of lambdoidal wormian bones ($N = 19$) and lateral coronal ossicles ($N = 8$). There is also a considerable degree of ectocranial porosity, which is common to most prehistoric Easter Islanders.

The other two crania identified by the multivariate analysis do not show the pattern of traits suggesting Caucasoid admixture exhibited by the previous skull. On the contrary, alternative diagnostic approaches produce strong Polynesian results.

The simometer scores for RH0064, a male aged 25 to 29 years from the site of Ahu Nau Nau, show a strong nonwhite result with two indices (MF, ZY) falling well into the range of nonwhite, and only the alpha index being intermediate. The large, long cranium (76.6 cranial index) is typical of Easter Islanders, as are the numerous lambdoidal wormian bones, an elliptic palate (almost never found in whites), and an obliterated coronal "three" suture line (common among Easter Island natives). The orbital form is elliptic, which is rare among whites. In total, about six traits typical of Polynesian/Easter Island crania are observable on this skull, but none of the traits common to whites is apparent.

The simometer results for AMNH No. 9819, a male aged 18 to 20 years, are strongly

Fig. 4. Distributions of discriminant scores for Easter Island and Caucasian males (top) and females (bottom).

nonwhite, with none of the values approaching the sectioning points. The discrete morphological traits show a clear non-Caucasoid pattern. The zygomaticomaxillary suture is "angled," the palatine suture is straight, the nasal sill is completely "guttered," and the nasal profile is concavo-convex. These traits are rare in whites; in combination, they strongly suggest that this individual does not have mixed parentage.

Thus, from midfacial metrics and diagnostic patterns of discrete cranial traits, RH0064 and AMNH No. 9819, quite unlike skull 29746, classify as eastern Polynesians. Skull 29746 presents anthroposcopic and morphometric patterns reflecting both Rapa Nui and Caucasian characteristics, and they are best explained as the result of admixture involving a European father and an island mother.

IMPLICATIONS OF CONTACT IN EASTER ISLANDERS

Rapa Nui was discovered by Europeans during the first quarter of the eighteenth century, two centuries after the discovery of the Americas. Unlike many islands, including Hawaii, Easter Island at the time of discovery had begun a downward spiral of cultural disintegration (Kirch, 1984). Endemic warfare resulted from overpopulation and environmental degradation.

The impression given by folklore and sporadic historical documentation is of chronic, lethal warfare during the late prehistoric and early historic periods. Based on the osteological evidence, this assessment is somewhat misleading. Fractures indicative of cranial trauma are fairly common, and examples of fatal injuries are evident; however, most skeletal injuries appear to have been nonlethal. Few fatalities were directly attributable to violence. The physical evidence suggests that the frequency of warfare and lethal events was exaggerated in folklore, presumably because of its horrific results and importance in the daily lives of participants.

Early encounters with Europeans were brief and caused little osteologically identifiable impact on the island's population through the first half of the nineteenth century. A few encounters ended in violence; the skeletal collection includes one individual wounded by a gunshot. Sexual liaisons were common, and the osteological data provide evidence of offspring from such unions.

Studies of biological distance using cranial measurements have traditionally examined intergroup relationships and population origins based on the assumption that morphological similarity reflects genetic affinity. Relatively little research has explored heterogeneity within a given sample (e.g., Jantz, 1974; Key and Jantz, 1990; Byrd and Jantz, n.d.), particularly in relation to the identification of Europeans or their offspring in native cemetery samples following initial contact (Jantz, 1973). Recently, through the application of forensic anthropological techniques for race determination, Jantz and Owsley (n.d.) were able to identify a Euramerican buried in an early protohistoric American Indian cemetery. The possibility of mixed-parent offspring was tentatively suggested based on within-sample heterogeneity, with specific individuals showing greater similarity to European morphology in a multivariate discriminant analysis.

A similar technique was applied in this study of Easter Islanders with successful results. One skull, in particular, has morphological and metric characteristics strongly suggesting mixed parentage, an event predictable from the historical record. Methodologically, this identification is significant for it demonstrates the utility of this approach in detecting genetic admixture. It also shows the ambiguity that can accompany this type of analysis. Two other skulls were similarly identified by the craniofacial multivariate analysis as white, one very strongly so, but alternative classification procedures did not support these assessments. Although these may be simple cases of misclassification, they could reflect how little we know about the heritability of skeletal characteristics in cases of mixed parentage.

Europeans also brought diseases to the island, with devastating results. Sexual contact led to the early transmission of venereal disease. Although examples of caries sicca were not identified, several long bones show subperiosteal changes that suggest venereal syphilis. Historical documents indicate that smallpox arrived more recently, later than the human remains observed in this study, but, as elsewhere, its introduction was catastrophic. Smallpox, tuberculosis, and turmoil caused by internal social upheaval and external strife decimated the native population during the last half of the nineteenth century.

The eventual result of discovery by Europeans was the near-complete extermination of the population and effective loss of their culture. History repeated itself in this result, although the circumstances, timing, and sequence of events were unique.

ACKNOWLEDGMENTS

We gratefully acknowledge the members of the Consejo de Monumentos, the Director of the Dirección de Bibliotecas, Archivos y Museos, and the Governor of Easter Island for their endorsement of the osteological research. Cooperation of the Easter Island Consejo de Ancianos is greatly appreciated. Claudio Cristino, Sergio Rapu H., Andrea Seelenfreund, and Leslie Shaw were responsible for field excavations of many of the archaeological sites. Eugenio Aspillaga, Mario Castro, Silvia Quevedo K., José Miguel Ramirez, Jaymie Brauer, and Fermina Zamorano provided assistance that facilitated examinations. Scott Baker, Carmen Clayton, Suzanne Bennett, Sonia Haoa C., and Robert Mann helped collect data. Jo Anne Van Tilburg and Tess Freidenburg provided historical documentation. The cranial measurements of historic period whites were provided by Richard Jantz. The collections examined are curated by the American Museum of Natural History (AMNH), New York; the Field Museum, Chicago; the Smithsonian Institution, Washington, DC; the National Museum of Natural History, Santiago, Chile; the Fonck Museum, Vina del Mar, Chile; the Natural History Museum, Valparaiso, Chile; and the Sabastion Englert Museum, Easter Island. We thank the curatorial staffs of these museums for allowing access to their facilities and collections. Grants enabling data collection were provided by the Government of Chile, the Center for Field Research-Earthwatch, the National Geographic Society, the University of Wyoming, the Smithsonian Institution, the Kon-Tiki Museum, Oslo, and the Dunn Estate Fund.

REFERENCES CITED

Ayres WS (n.d.) Radiocarbon dates from Easter Island. Manuscript on file, Department of Anthropology, University of Oregon, Eugene.

Bass WM (1987) Human Osteology: A Laboratory and Field Manual of the Human Skeleton, 3rd ed. Columbia: Missouri Archaeological Society.

Belcher E (1825–1827) Pacific journal and remarks. Pacific Manuscript Bureau (PMB), ANU, microfilm, M51.

Boutilier JA (1992) METEI: A Canadian medical expedition to Easter Island, 1964–65. Rapa Nui Journal 6:21–22, 26–33.

Brues AM (1990) The once & future diagnosis of race. In G Gill and S Rhine (eds.): Skeletal Attribution of Race: Methods for Forensic Anthropology. Maxwell Museum of Anthropology, Anthropological Papers 4:1–7.

Byrd JE, and Jantz RL (n.d.) Osteological evidence for distinct social groups at the Leavenworth site (39CO9). In DW Owsley and RL Jantz (eds.): Skeletal Biology in the Great Plains. Washington, DC: Smithsonian Institution Press.

Campbell NA (1984) Some aspects of allocation and discrimination. In GN Van Vark and WW Howells (eds.): Multivariate Statistical Methods in Physical Anthropology. Dordrecht: D. Reidel Publishing Co., pp. 177–192.

El-Najjar MY, and McWilliams KR (1978) Forensic Anthropology: The Structure, Morphology, and Variation of Human Bone and Dentition. Springfield: Charles C. Thomas.

Englert S (1970) Island at the Center of the World. New York: Charles Scribner's Sons.

Forster G (1777) A Voyage Round the World in His Britannic Majesty's Sloop, Resolution, Commanded by Capt. James Cook, During the Years 1772, 3, 4, and 5, vol. I. London: printed for B. White, Fleet-Street.

Forster JR (1982) The Resolution Journal of Johann Reinhold Forster 1772–1775, vol. III. London: The Hakluyt Society.

Gill GW (1971) The Prehistoric Inhabitants of Northern Coastal Nayarit: Skeletal Analysis and Description of Burials. Ph.D. dissertation, Department of Anthropology, University of Kansas, Lawrence.

Gill GW (1984) A forensic test case for a new method of geographical race determination. In TA Rathbun and JE Buikstra (eds.): Human Identification: Case Studies in Forensic Anthropology. Springfield: Charles C. Thomas, pp. 329–339.

Gill GW (1986a) Final Report of Investigation of the 1981 Easter Island Anthropological Expedition. Report to the National Geographic Society, grant no. 2256–80.

Gill GW (1986b) Craniofacial criteria in forensic race identification. In KJ Reichs (ed.): Forensic Osteology: Advances in the Identification of Human Remains. Springfield: Charles C. Thomas, pp. 143–159.

Gill GW, Hughes SS, Bennett SM, and Gilbert BM (1988) Racial identification from the midfacial skeleton with special reference to American Indians and whites. Journal of Forensic Sciences *33*:92–99.

Gill GW, Owsley DW, and Baker SJ (1983) Craniometric evaluation of prehistoric Easter Island populations. American Journal of Physical Anthropology *60*:197.

Gill GW, and Owsley DW (1993) Human osteology of Rapanui. In SR Fischer (ed.): Rapanui Studies: Essays in Memory of William T. Mulloy. Oxford: Oxbow Monograph 32, pp. 56–62.

Gonnell FA (1992) The slave trade run by the Spaniards in the Pacific Ocean during the 19th century: The case of Easter Island. Rapa Nui Journal *6*:23–25.

Hackett CJ (1976) Diagnostic Criteria of Syphilis, Yaws and Treponarid (Treponematoses) and of Some Other Diseases in Dry Bones. Berlin: Springer–Verlag.

Hamilton JA (1856) Massacre at Easter Island. The Friend *13*:50.

Houghton P (1977) Rocker jaws. American Journal of Physical Anthropology *47*:365–370.

Howells WW (1973) Cranial Variation in Man. Papers of the Peabody Museum of Archaeology and Ethnology, Harvard University 67.

Jantz RL (1970) Change and Variation in Skeletal Populations of Arikara Indians. Ph.D. dissertation, Department of Anthropology, University of Kansas, Lawrence.

Jantz RL (1973) Microevolutionary change in Arikara crania: A multivariate analysis. American Journal of Physical Anthropology *38*:15–26.

Jantz RL (1974) The Redbird focus: Cranial evidence in tribal identification. Plains Anthropologist *19*:5–13.

Jantz RL, and Owsley DW (n.d.) White traders in the upper Missouri: Osteological evidence from the Swan Creek site. In DW Owsley and RL Jantz (eds.): Skeletal Biology in the Great Plains. Washington, DC: Smithsonian Institution Press.

Key PJ, and Jantz RL (1990) Statistical assessment of population variability: A methodological approach. American Journal of Physical Anthropology. *82*:53–59.

Kirch PV (1984) The Evolution of Polynesian Chiefdoms. Cambridge: Cambridge University Press.

La Perouse JF (1968) A Voyage Round the World Performed in the Years 1785, 1786, 1787 and 1788 by the Boussole and Astrolabe, vol. I. New York: Da Capo Press.

Mann RW, and Murphy SP (1990) Regional Atlas of Bone Disease: A Guide to Pathologic and Normal Variation in the Human Skeleton. Springfield: Charles C. Thomas.

McCall G (1976) European impact on Easter Island: Response, recruitment and the Polynesian experience in Peru. Journal of Pacific History *11*:90–105.

McCall G (1981) Rapanui Tradition and Survival on Easter Island. Honolulu: University Press of Hawaii.

McCoy PC (1976) Easter Island Settlement Patterns in the Late Prehistoric and Protohistoric Periods. New York: International Fund for Monuments Bulletin 5.

Metraux A (1940) Ethnology of Easter Island. Bernice P. Bishop Museum Bulletin 160.

Metraux A (1957) Easter Island. New York: Oxford University Press.

Mulloy W (1988) A preliminary culture-historical research model for Easter Island. Rapa Nui Notes *7*:2–4.

Murrill RI (1968) Cranial and Postcranial Skeletal Remains from Easter Island. Minneapolis: University of Minnesota Press.

Ortner DJ, and Putschar WGJ (1981) Identification of Pathological Conditions in Human Skeletal Remains. Smithsonian Contributions to Anthropology 28.

Porteous JD (1981) The Modernization of Easter Island. Western Geographical Series 19.

Routledge KP (1919) The Mystery of Easter Island. London: Sifton, Praed & Co. Ltd.

SAS Institute Inc. (1985) SAS User's Guide: Statistics. 5th ed. Cary: SAS Institute Inc.

Van Tilburg J (1992) HMS Topaze on Easter Island: Hoa Hakananai'a and Five Other Museum Sculptures in Archaeological Context. British Museum Occasional Paper 73.

von Saher H (1992) More journals on Easter Island: The works of Johann Reinhold Forster (1729–1798) and Johann George Adam Forster (1754–1794) [Part II]. Rapa Nui Journal 6: 34–39.

Zimmerman MR, and Kelley MA (1982) Atlas of Human Paleopathology. New York: Praeger.

Chapter 15

An Osteological Assessment of Health and Disease in Precontact and Historic (1778) Hawai'i

Michael Pietrusewsky and Michele T. Douglas

Department of Anthropology, University of Hawai'i at Mānoa, Honolulu, HI 96822

Nearly three centuries separate Christopher Columbus's landing in the New World in 1492 and Captain James Cook's first sighting of the Hawaiian Islands on January 18, 1778. Although later in time and in a different ocean, the consequences of this subsequent "discovery" are comparable to Columbus's discovery of the New World. Both achievements altered, often calamitously, the lives and destinies of both the indigenous people living in these newly "discovered" places as well as those living in the Western world from which these early explorers sailed. In Hawai'i, European explorers were followed by merchants, missionaries, and foreign (especially Asian) laborers who gave rise to the multiracial population now living there. With contact came new diseases, technology, and ideas that inevitably led to a disruption of the traditional culture of the newly contacted people. The introduction of new infectious diseases, coupled with lower fertility, caused the drastic decimation of entire indigenous populations.

This chapter documents biological change in Hawai'i at and following contact using a variety of skeletal and dental indicators. It is the first such synthesis comparing precontact and contact health and disease in Hawai'i. Using published and unpublished osteological data, these comparisons provide an opportunity to assess Hawaiian health before, at, and after Western contact. The latter is traditionally identified as beginning with Captain Cook's first sighting of the Hawaiian Islands in 1778. Among the issues that we address is the presence or absence of infectious diseases such as tuberculosis and treponemal infection in precontact Hawai'i, diseases that leave a recognizable imprint on the skeleton. This brief survey of health and disease in precontact and contact Hawaiians provides a base for future studies in this area.

EARLY HAWAIIANS

Attesting to its extreme geographic isolation in the Pacific, the first people to arrive in the Hawaiian Islands around the fourth and fifth centuries A.D. were Polynesians sailing in large, double-hulled voyaging canoes from the Marquesas Islands in eastern Polynesia (Kirch, 1985:1). From a small founding population, estimated to

In the Wake of Contact: Biological Responses to Conquest, pages 179–196

have been only a few hundred people, with an economy based on agriculture and fishing, there emerged one of the most complex and sophisticated cultures of the Pacific (Kirch, 1985). At the time of European contact, a strongly hierarchical society composed of independent chiefdoms, conservatively estimated at about 250,000 people, was in the process of islandwide unification (Kirch, 1984, 1985).

Following contact, however, the native Hawaiian population collapsed so precipitously that by the time of the Great Mahele in 1848, when land was redistributed to private individuals, the number of Hawaiians was estimated to have been only about 80,000. By any measure, the decline in numbers was great, but by using Stannard's (1989) more generous estimate of 800,000 or more Hawaiians at contact, it was truly catastrophic.

According to native scholar Kamakau (1964:98), high-mortality contagious diseases like measles and smallpox, along with venereal infections, leprosy, and yaws, were unknown prior to European contact. Partial validation of this viewpoint is provided by Cook and his crew who, after a 10-month interval between their first and second visits, reported that venereal disease (syphilis and gonorrhea) had become widespread among the population (Beaglehole, 1967). Epidemics recorded as sweeping through Hawaii's postcontact population include measles, whopping cough, influenza, smallpox, scarlet fever, typhoid fever, Asiatic cholera, bubonic plague, and yellow fever (Halford, 1954). Other introduced diseases included tuberculosis, leprosy, mumps, dysentery, and typhus.

SKELETAL REMAINS

The earliest studies of Hawaiian osteology (e.g., Uhde, 1861; Davis, 1867; Allen, 1898; and others) focused on craniometry, typological racial classifications, and theories of migration. These reports provide little, if any, direct information on issues such as health, demography, and quality of life. Later studies benefited from some of the first, albeit primitive, excavations of ancient Hawaiian cemeteries and provided our first glimpses of the biological and demographic characteristics of early Hawaiians. One of the first major osteological studies was by Snow (1974) who examined over 1,000 skeletons from Mōkapu on O'ahu. Other published studies, usually based on far fewer individuals, include Underwood (1969) for Pu'u Ali'i at South Point, Pietrusewsky (1971) for 'Anaeho'omalu, Kam (1979) for Kalāhuipua'a, and Collins (1986) for Keōpū, all from the island of Hawai'i.

Clearly, the database for assessing the disease experience of precontact and contact period Hawaiians, while uneven, has substantially improved in recent years. Comparative information on mortality, adult stature, enamel hypoplasia, cribra orbitalia, specific and nonspecific infectious diseases, trauma, degenerative conditions, and dental lesions are increasingly available. Other potential indicators of health, including Harris lines, fluctuating or random asymmetry, enamel microdefects, isotopic and trace element studies, have rarely, if ever, been systematically recorded in Hawaiian remains. Consenting to the wishes of some native Hawaiian activist groups, the use of invasive techniques, radiography, or even photography are, for the most part, no longer allowed in bioarchaeological research.

Determining the temporal status of skeletal series from Hawai'i will always be problematical. We have tried to select those series whose temporal provenience is generally well known. However, the intervening "historic" period from initial contact (1778) to the present is a relatively brief time interval and is difficult to identify archaeologically in the poorly stratified rock and sand typical of Hawaiian sites. Traditionally, the absence of historical burial goods and the manner of interment (e.g., flexed rather than supine) have been the markers used to determine the precontact status of human remains. However, a 40-year period separates Captain Cook's arrival in 1778 and the establishment of trade and missionary influence (1820) when cultural and material traditions underwent widespread changes (Carter, 1990). Thus, a window was created where the burials appear prehistoric but are actually from the historic/contact period. The absence of adequate dates for many existing skeletal series and current prohibitions against obtaining dates from human bone only confound the dating problem.

Methodological differences in recording and reporting frequencies limit the use of other researchers' data, although in Hawai'i we are fortunate that workers are increasingly adopting a common standard for collecting osteological information. Because of incompatibility problems, we emphasize personally recorded data. The skeletal series include seven prehistoric and four historic samples from four of the largest and most populous islands in the Hawaiian chain (Fig. 1, Table 1). The historic samples are typically accumulated over a shorter time period than the prehistoric samples.

Two of the series, Wai'anae and Keōpū, each contain precontact and contact specimens that have been separated for analysis. The 'Anaeho'omalu series represents late precontact and early contact specimens that could not be segregated and, for that reason, they have not been included in either the precontact or postcontact totals presented here. Differences in precontact versus postcontact indicators of health and disease were tested using the chi-square statistic (χ^2) and, where applicable, Yates' correction for continuity (Thomas, 1986).

THE OSTEOLOGICAL PICTURE

Mortality Data

Although paleodemographic estimates are not without their critics (e.g., Bocquet-Appel and Masset, 1982; Howell, 1982; Sattenspiel and Harpending, 1983) age-at-death data hold promise for understanding past populations. Unfortunately, because of uneven and small sample sizes, paleodemographic data for contact period Hawai'i is insufficient for an in-depth discussion of changes between the precontact and postcontact periods. Further, the underrepresentation of subadults in many series means that researchers must rely heavily on the mean age-at-death of adults in comparisons of skeletal collections.

The proportion of deaths in individuals less than 20 years of age is a useful measure of the degree of underrepresentation and overrepresentation of subadults in skeletal samples and, in optimal situations, a gauge of childhood mortality. Assuming that the proportion of dead under 15 years of age in a stationary population should lie between 30% and 50% (Weiss, 1973:49), subadults are underrepresented in all precontact series from Hawai'i except Honokahua (Table 2). This discrepancy is even more apparent in the historic series.

The sex ratio (males:females) ranges from 71.4 to 120 in the precontact series (lower scores indicate male underrepresentation). The overrepresentation of adults to subadults (88 versus 12) and the disproportionately higher male to female ratio in the 'Anaeho'omalu series suggest that these burial caves were not the result of a long-term habitation area such as a village, but the result of differential use of the site by adult males harvesting land and sea resources (Pietrusewsky et al., 1990b). A disproportionate number of males is further evident in the historic Kaka'ako series, but since the entire cemetery was not excavated, this finding may be due to sampling error.

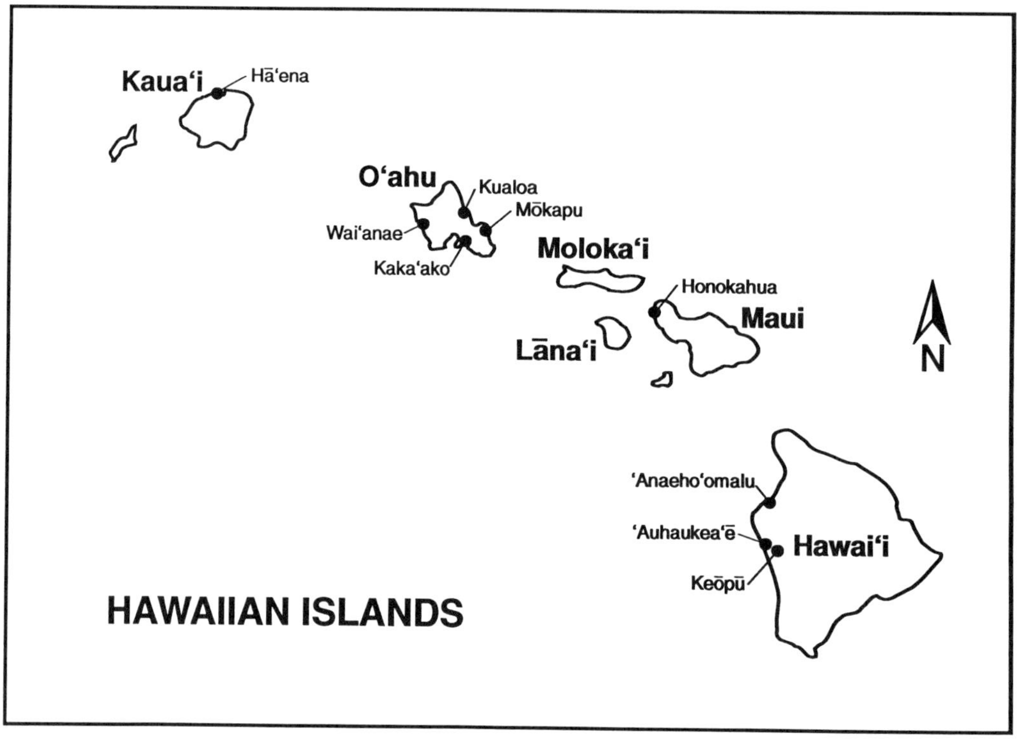

Fig. 1. Map of the Hawaiian Islands showing the approximate location of the skeletal series used in the present survey.

Underrepresentation of males at most sites in Hawai'i has been suggested to be the result of differential burial practices for males, such as burial at sites of differential use, burial at sea, or during interisland warfare, burial at the battleground (Pietrusewsky et al., 1991).

The mean age-at-death for the historic (34.3–42.4 years) and prehistoric (24.5–47.6 years) samples does not reveal any clear temporal pattern. The lowest value (24.5 years) occurs in the Mokapu series and reflects the limitations of the age-estimation methods in use at the time of analysis, while the sample from Honokahua was examined using the latest in estimation techniques. The mean age-at-death is lower in the samples with an overrepresentation of males (Kualoa and 'Anaeho'omalu).

Stature

Because no regression formulas specific for Hawaiian skeletal remains are available, stature estimates are generally made using New Zealand Maori (Polynesian) regression formulas (Houghton et al., 1974). The mean average stature in prehistoric Hawaiian males (172.4 cm) is only slightly less than that determined for historic males (173.5 cm; Table 3). Hawaiian males measured in 1920–1921 were slightly shorter (169.5 cm) than the skeletal series. Historic females, on average (162.9 cm), are a half inch taller than precontact Hawaiian females (161.6 cm). Hawaiian females measured in 1920 were very close to both of these estimates (159 cm).

Enamel Hypoplasia

Enamel hypoplasia results from a disruption of enamel production in the growing

TABLE 1. Hawaiian Skeletal Series Used in this Study

Skeletal series	Location	Size	Dates	References
Historic				
Kaka‘ako	O‘ahu	28	1853–1854	Pietrusewsky et al., 1989
Wai‘anae	O‘ahu	19	1865–1900	Pietrusewsky and Douglas, 1990
Auhaukea‘e	Hawai‘i	24	1890–1906	Pietrusewsky and Douglas, 1989b
Keōpū	Hawai‘i	13	1840–1930	Collins, 1986
Historic/prehistoric				
‘Anaeho‘omalu	Hawai‘i	100	Prehistoric/historic	Pietrusewsky, 1971; Pietrusewsky et al., 1990b
Prehistoric				
Mōkapu	O‘ahu	1,171	Prehistoric	Snow, 1974
Kualoa	O‘ahu	42	A.D. 1400–1600	Pietrusewsky and Douglas, 1989a
Wai‘anae	O‘ahu	14	Prehistoric	Pietrusewsky and Douglas, 1990
Hā‘ena	Kaua‘i	31	Prehistoric	Pietrusewsky et al., 1992
Honokahua	Maui	712	A.D. 610–1800	Pietrusewsky et al., 1991
Keōpū	Hawai‘i	342	A.D. 1330–1780	Collins, 1986

TABLE 2. A Comparison of Some Demographic Features of Hawaiian Skeletal Series[a]

Skeletal series	No. of subadults		No. of adults		No. of males	No. of females	Adult sex ratio[b]	Mean age-at-death[c]
	N	%	N	%				
Historic								
Kaka'ako, O'ahu	2	9.1	20	90.9	11	9	122.2	34.3
Auhaukea'e, Hawai'i	2	11.1	16	88.9	6	9	66.7	42.4
Wai'anae, O'ahu	1	20.0	4	80.0	1	3	33.3	39.4
Historic total	5	11.1	40	88.9	18	21	85.7	
Historic/prehistoric								
'Anaeho'omalu, Hawai'i	6	6.0	94	94.0	52	36	144.4	30.6
Prehistoric								
Mōkapu, O'ahu	307	26.8	838	73.2	398	439	90.7	24.5[d]
Kualoa, O'ahu	10	23.8	31	75.6	16	14	114.3	33.7
Wai'anae, O'ahu	1	7.1	13	92.9	5	7	71.4	47.6
Hāena, Kaua'i	9	29.0	27	71.0	12	10	120.0	38.3
Honokahua, Maui	278	38.9	434	61.1	143	255	56.1	39.1
Keōpū, Hawai'i	75	22.1	265	77.9	113	130	86.9	34.1
Prehistoric total	680	29.7	1,608	70.3	687	855	80.4	

[a]See Table 1 for references.
[b]Computed by the formula: $(M \div F) \times 100$.
[c]In adults (> 20 years).
[d]This figure is taken from the life table constructed by Collins (1986:242).

TABLE 3. Comparison of Stature (in cm) in Historic and Prehistoric Hawaiians[a]

Skeletal samples/sex	N	Average	Minimum	Maximum
Historic samples—males				
Kaka'ako, O'ahu	10	173.4	165.9	178.4
Auhaukea'e, Hawai'i	4	173.9	170.8	175.8
Wai'anae, O'ahu	1	170.3		
Keōpū, Hawai'i	7	176.5	173.0	182.4
Hawaiians ca. 1920[b]	204	169.5		
Prehistoric samples—males				
'Anaeho'omalu, Hawai'i	40	173.5	162.1	179.6
Keōpū, Hawai'i	31	172.0	162.3	176.8
Honokahua, Maui	100	173.3	164.7	182.0
Hā'ena, Kaua'i	10	171.8	168.0	176.5
Mōkapu, O'ahu	75	172.9	162.1	181.5
Kualoa, O'ahu	9	171.6	166.7	180.5
Wai'anae, O'ahu	4	173.1	167.3	179.4
Historic samples—females				
Kaka'ako, O'ahu	8	162.4	160.1	165.2
Auhaukea'e, Hawai'i	7	158.3	165.9	163.6
Wai'anae, O'ahu	3	164.5	159.1	169.4
Keōpū, Hawai'i	5	166.2	159.4	177.9
Hawaiians ca. 1920	173	159.0		
Prehistoric samples—females				
'Anaeho'omalu, Hawai'i	27	162.1	147.0	169.0
Keōpū, Hawai'i	45	161.8	142.4	170.3
Honokahua, Maui	221	163.1	149.8	172.0
Hā'ena, Kaua'i	8	162.0	157.0	167.8
Mōkapu, O'ahu	115	158.1	145.0	175.0
Kualoa, O'ahu	9	162.5	158.3	165.4
Wai'anae, O'ahu	4	162.0	159.7	164.0

[a]See Table 1 for references.
[b]Sullivan (1927).

deciduous or permanent tooth bud, most commonly the incisors and canines, and it is caused by a variety of nutritional, emotional, and infectious disease stresses (Goodman et al., 1980; Lukacs, 1989). Enamel hypoplasia, here reported for adult male and female Hawaiian incisors and canines (Table 4), is more common (23.8%) in teeth from historic skeletons than in those from precontact times (7.7%), and the difference is statistically significant (χ^2, $p \leq 0.05$). These comparative data suggest that these indicators of general stress occurring at early ages were more common in contact times than in prehistory. While there is also considerable variance in the prehistoric series, this may be attributed to the problem of defining "precontact." Both the 'Anaeho'omalu and Wai'anae sites are transitional samples.

Cribra Orbitalia

Porotic hyperostosis and cribra orbitalia are commonly attributed to anemias, including iron-deficiency anemia (Steinbock, 1976; Stuart-Macadam, 1987). No significant difference in the prevalence of cribra orbitalia in precontact (13.5%) and postcontact (25.0%) adults was observed (Table 5). Cribra orbitalia is more frequent in subadult Hawaiians, 42.5% in prehistoric and 33.3% in historic skeletons, but the dif-

TABLE 4. Frequency of Occurrence of Enamel Dental Hypoplasia[a] in Hawaiian Remains

Skeletal series	A/O[b]	%
Historic		
Kaka'ako, O'ahu	14/99	14.1
Auhaukea'e, Hawai'i	25/93	26.9
Wai'anae, O'ahu	14/31	45.2
Historic total	53/223	23.8
Historic/prehistoric		
'Anaeho'omalu, Hawai'i	19/80	23.8
Prehistoric		
Honokahua, Maui	102/1,718	5.9
Hā'ena, Kaua'i	24/181	13.3
Kualoa, O'ahu	21/119	17.6
Wai'anae, O'ahu	12/35	34.3
Prehistoric total	159/2,053	7.7

[a]Any expression of hypoplastic defect in canines and incisors.
[b]A/O, affected/observed.

ference between these two groups is not significant. Despite an apparent higher frequency of cribra orbitalia in the historic series, tests of significance do not support a general increase in this indicator of ill health from precontact to postcontact times in Hawai'i.

Limb Bone Fractures

Limb bone fractures provide an indication of the frequency and type of accidental or deliberate injuries in a population (Walker, 1989). The frequencies of bone fractures in Hawaiian skeletal remains were determined using the corresponding numbers of complete skeletal elements (Table 6). Skull fractures and fractures of other bones not reported in this table occur in Hawaiian remains, but the problems of systematically reporting their occurrence in complete specimens are formidable.

The combined frequency of limb bone fractures is 1.6% in the precontact series and 1.0% in the historic series; the difference is not statistically significant (χ^2, $p > 0.05$). Radii and ulnae are the most frequently fractured bones in the precontact series that consist of 6,555 bones, which is large by archaeological standards. At Honokahua, the largest precontact skeletal series examined in recent years, the most common fracture is of the distal ulna, also known as the "parry" fracture, because it is sustained when the arm is raised to protect the head against a blow. The left ulna is more commonly fractured (68.8%) than the right ulna (31.3%) in this same series (Pietrusewsky et al., 1991). Four of the 16 distal ulna fractures were in males and 12 occur in 11 females (one female has two ulnae fractures) from this same site. The occurrence of ulnae fractures in both males and females at Honokahua is suggestive of interpersonal violence within the household rather than in warfare. Available weapons of war in ancient Hawai'i included slingstones, spears, daggers, and clubs, many of which could result in a lethal injury to the soft tissue with no evidence in the skeletal remains. However, the absence of consistent patterns of fractures in Hawaiian skeletal series, in the presence of ethnographic evidence of warfare, is enigmatic or is evidence of differential burial practices.

Degenerative Articular Changes

Degenerative osteoarthritis on the articular joint surfaces of Hawaiian remains reflects for the most part biomechanical wear and tear and functional stress (e.g., Goodman et al., 1980). Although initially scored on a scale of none, slight, medium, and marked arthritis, these frequencies only reflect moderate or marked expressions of lipping or porosis.

Statistical tests of the frequencies of osteoarthritis in each joint surface reveal significant differences between the historic and prehistoric samples in the sternoclavicular, acromioclavicular and humeral head (χ^2, $p \leqslant 0.05$). The historic samples have a higher frequency of osteoarthritis of the sternoclavicular and humeral head ar-

TABLE 5. Frequency of Occurrence of Cribra Orbitalia in Hawaiian Skeletal Series (Reported by Individual)[a]

	Adult males		Adult females		Adults		Subadults		Total	
Skeletal series	A/O	%	A/O	%	A/O	%	A/O	%	A/O	%
Historic										
Kaka‘ako, O‘ahu	1/8	12.5	1/5	20.0	2/13	15.4	1/1	100.0	3/14	21.4
Auhaukea‘e, Hawai‘i	0/3	0.0	1/4	25.0	1/7	14.3	0/1	0.0	1/8	12.5
Wai‘anae, O‘ahu	0/1	0.0	3/3	100.0	3/4	75.0	0/1	0.0	3/5	60.0
Historic total	1/12	8.3	5/12	41.7	6/24	25.0	1/3	33.3	7/27	25.9
Historic/prehistoric										
‘Anaeho‘omalu, Hawai‘i	1/20	5.0	0/20	0.0	1/40	2.5	1/2	50.0	2/42	4.8
Prehistoric										
Keōpū, Hawai‘i	5/41	12.2	4/27	14.8	9/68	13.2	—	—	9/68	13.2
Honokahua, Maui	8/100	8.0	11/212	5.2	19/312	6.1	73/155	47.1	92/468	19.7
Hā‘ena, Kaua‘i	0/6	0.0	3/11	27.3	3/17	17.6	5/6	83.3	8/23	34.8
Mōkapu, O‘ahu[b]	20/154	13.0	18/149	12.1	38/303	12.5	—	—	38/303	12.5
Mōkapu, O‘ahu	24/132	18.2	42/176	23.9	66/308	21.4	52/148	35.1	118/456	25.9
Kualoa, O‘ahu	1/7	14.3	1/10	10.0	2/17	11.8	3/4	75.0	5/22	22.7
Wai‘anae, O‘ahu	2/3	66.7	0/3	0.0	2/6	33.3	0/0	0.0	2/6	33.3
Prehistoric total	60/443	13.5	79/588	13.4	139/1,031	13.5	133/313	42.5	272/1,346	20.2

[a]See Table 1 for references.
[b]Suzuki (1987).

TABLE 6. Limb Bone Fractures in Adult Hawaiian Remains[a]

	Clavicle		Humerus		Radius		Ulna	
Skeletal sample	A/O	%	A/O	%	A/O	%	A/O	%
Historic								
Kaka‘ako, O‘ahu	0/14	0.0	0/13	0.0	0/12	0.0	0/13	0.0
Wai‘anae, O‘ahu	0/8	0.0	1/7	14.3	0/6	0.0	0/6	0.0
Auhaukea‘e, Hawai‘i	0/11	0.0	0/11	0.0	1/10	10.0	0/5	0.0
Historic total	0/33	0.0	1/31	3.2	1/28	3.6	0/24	0.0
Historic/prehistoric								
Anaeho‘omalu, Hawai‘i	1/33	3.0	0/74	0.0	1/66	1.5	3/62	4.8
Prehistoric								
Mōkapu, O‘ahu	1/342	0.3	6/438	1.4	7/376	1.9	9/438	2.1
Kualoa, O‘ahu	0/4	0.0	1/6	16.7	0/9	0.0	0/5	0.0
Hā‘ena, Kaua‘i	0/17	0.0	1/20	5.0	1/19	5.3	0/19	0.0
Honokahua, Maui	6/304	2.0	4/442	0.9	10/476	2.1	22/406	5.4
Keōpū, Hawai‘i	3/51	5.9	0/100	0.0	2/92	2.2	3/70	4.3
Wai‘anae, O‘ahu	0/9	0.0	0/9	0.0	0/10	0.0	0/9	0.0
Prehistoric total	10/727	1.4	12/1,015	1.2	20/982	2.0	34/947	3.6

	Femur		Tibia		Fibula		Total	
	A/O	%	A/O	%	A/O	%	A/O	%
Historic								
Kaka‘ako, O‘ahu	0/10	0.0	0/12	0.0	0/11	0.0	0/85	0.0
Wai‘anae, O‘ahu	0/6	0.0	0/6	0.0	0/5	0.0	1/44	2.3
Auhaukea‘e, Hawai‘i	0/13	0.0	0/9	0.0	0/7	0.0	1/66	1.5
Historic total	0/29	0.0	0/27	0.0	0/23	0.0	2/195	1.0
Historic/prehistoric								
Anaeho‘omalu, Hawai‘i	0/86	0.0	0/73	0.0	0/59	0.0	5/453	1.1
Prehistoric								
Mōkapu, O‘ahu	2/469	0.4	5/441	1.1	7/280	2.5	37/2,784	1.3
Kualoa, O‘ahu	0/5	0.0	0/0	0.0	0/2	0.0	1/31	3.2
Hā‘ena, Kaua‘i	0/20	0.0	0/20	0.0	0/16	0.0	2/131	1.5
Honokahua, Maui	5/491	1.0	3/465	0.6	7/365	1.9	57/2,949	1.9
Keōpū, Hawai‘i	2/113	1.8	0/103	0.0	0/67	0.0	10/596	1.7
Wai‘anae, O‘ahu	0/9	0.0	0/8	0.0	0/10	0.0	0/64	0.0
Prehistoric total	9/1,107	0.8	8/1,037	0.8	14/740	1.9	107/6,555	1.6

[a]See Table 1 for references.

ticular surfaces while the prehistoric samples have a higher frequency of osteoarthritis of the acromioclavicular surface. Although these results were mixed, they do suggest a possible alternative use of the upper arm in the historic skeletal samples. Combining the total frequency of advanced osteoarthritis in several articular joint surfaces (Table 7) results in an average frequency of 3.9% in the precontact series and 7.3% in the historic series, a statistically significant difference (χ^2, $p \leq 0.01$).

Very similar levels of advanced (moderate to marked) lipping are observed in the vertebral centra of precontact (9.1%) and historic (9.3%) series (Table 8). Osteoarthritic lipping of the articular surfaces in precontact vertebrae (6.1%) again matches the historic (5.9%) series. These differences were not found to be statistically significant (χ^2, $p > 0.05$).

TABLE 7. Frequency of Occurrence of Advanced Osteoarthritis in Hawaiian Skeletal Series[a]

	Historic							
	Kaka'ako		Wai'anae		Auhaukea'e		Total	
Articular surface	A/O	%	A/O	%	A/O	%	A/O	%
Sternoclavicular	4/17	23.5	1/8	12.5	2/14	14.3	7/39	17.9
Acromioclavicular	2/17	11.8	2/7	28.6	0/11	0.0	4/35	11.4
Glenoid fossa	0/20	0.0	2/8	25.0	0/16	0.0	2/44	4.5
Humeral head	6/16	37.5	0/7	0.0	0/11	0.0	6/35	17.1
Radial head	1/14	7.1	0/6	0.0	0/13	0.0	1/33	3.0
Proximal ulna	1/17	5.9	2/8	25.0	2/15	13.3	5/40	12.5
Sacroiliac	0/15	0.0	2/6	33.3	0/11	0.0	2/17	11.8
Acetabulum	0/16	0.0	1/6	16.7	0/14	0.0	1/36	2.8
Femoral head	0/12	0.0	0/6	0.0	0/17	0.0	0/35	0.0
Femoral condyle	0/15	0.0	2/6	33.3	2/18	11.1	4/39	10.3
Proximal tibia	0/17	0.0	1/6	16.7	2/17	11.8	3/40	7.5
Calcaneus	0/22	0.0	1/6	16.7	0/9	0.0	1/37	2.7
Talus	0/18	0.0	0/6	0.0	0/10	0.0	0/34	0.0
Temporomandibular	1/24	4.2	0/8	0.0	0/10	0.0	1/42	2.4
Total	15/240	6.3	14/94	14.9	8/186	4.3	37/506	7.3

	Historic/ prehistoric		Prehistoric							
	'Anaeho'omalu		Honokahua		Kualoa		Hā'ena		Total	
	A/O	%	A/O	%	A/O	%	A/O	%	A/O	%
Sternoclavicular	6/36	16.7	18/472	3.8	0/13	0.0	1/24	4.2	19/509	3.7
Acromioclavicular	9/34	26.5	8/421	1.9	0/6	0.0	0/15	0.0	8/442	18.1
Glenoid fossa	11/40	27.5	19/555	3.4	0/26	0.0	1/24	4.2	20/605	3.3
Humeral head	9/68	13.2	12/457	2.6	0/18	0.0	0/22	0.0	12/497	2.4
Radial head	4/59	6.8	6/487	1.2	0/19	0.0	0/32	0.0	6/538	1.1
Proximal ulna	10/80	12.5	39/536	7.3	1/24	4.2	0/32	0.0	40/592	6.8
Sacroiliac	8/47	17.0	39/535	7.3	0/13	0.0	3/25	12.0	42/573	7.3
Acetabulum	8/44	18.2	17/556	3.1	2/29	6.9	0/27	0.0	19/612	3.1
Femoral head	3/81	3.7	10/518	1.9	0/24	0.0	2/28	7.1	12/570	2.1
Femoral condyle	24/96	25.0	2/33	6.1	2/25	8.0	2/33	6.1	6/91	6.6
Proximal tibia	8/72	11.1	22/454	4.8	2/17	11.8	0/26	0.0	24/497	4.8
Calcaneus	2/56	3.6	21/571	3.7	1/22	4.5	0/33	0.0	22/626	3.5
Talus	3/62	4.8	23/580	4.0	1/26	3.8	0/35	0.0	24/641	3.7
Temporomandibular	0/33	0.0	18/221	8.1	3/17	17.6	1/12	8.3	22/250	8.8
Total	105/808	13.0	254/6,396	4.0	12/279	4.3	10/368	2.7	276/7,043	3.9

[a]See Table 1 for references.

Dental Pathology

Premortem tooth loss, dental caries, and abscessing in Hawaiian remains are presented on a per tooth or tooth socket basis (Table 9). The loss of teeth before death is typically indicative of pathological processes including periodontal disease, caries, and abscessing. The purposeful extraction of teeth (evulsion) as a sign of mourning, known to occur among Hawaiians (Pietrusewsky and Douglas, 1992) are not included in these tabulations.

Combining the premortem tooth loss frequencies indicates no significant difference

TABLE 8. Osteoarthritis and Osteophytosis in Hawaiian Vertebrae[a]

	Historic								Historic/prehistoric/	
	Kaka'ako		Auhaukea'e		Wai'anae		Total		'Anaeho'omalu	
Skeletal samples	A/O	%	A/O	%	A/O	%	A/O	%	A/O	%
Osteoarthritis of articular surfaces										
Cervical	0/542	0.0	2/176	1.1	14/110	12.7	14/625	2.2	25/472	5.3
Thoracic	20/375	5.3	5/295	1.7	20/166	12.0	40/507	7.9	77/849	9.1
Lumbar	10/366	2.7	5/111	4.5	27/53	50.9	37/419	8.8	103/410	25.1
Total	30/1,283	2.3	12/582	2.1	61/329	18.5	91/1,551	5.9	205/1,731	11.8
Osteophytosis of centra										
Cervical	4/104	3.8	3/67	4.5	6/46	13.0	10/139	7.2	17/189	9.0
Thoracic	5/201	2.5	17/159	10.7	20/72	27.8	25/269	9.3	61/416	14.7
Lumbar	3/81	3.7	18/64	28.1	11/36	30.6	14/117	12.0	57/210	27.1
Total	12/386	3.1	38/290	13.1	37/154	24.0	49/525	9.3	135/815	16.6

	Prehistoric							
	Honokahua		Kualoa		Ha'ena		Total	
	A/O	%	A/O	%	A/O	%	A/O	%
Osteoarthritis of articular surfaces								
Cervical	253/7,288	3.5	19/405	4.7	13/348	3.7	285/8,041	3.5
Thoracic	567/12,516	4.5	43/582	7.4	13/507	2.6	623/13,605	4.6
Lumbar	761/5,382	14.1	11/178	6.2	2/238	0.8	774/5,798	13.3
Total	1,581/25,186	6.3	73/1,165	6.3	28/1,093	2.6	1,682/27,444	6.1
Osteophytosis of centra								
Cervical	193/2,804	6.9	17/163	10.4	7/161	4.3	217/3,128	6.9
Thoracic	373/5,672	6.6	25/217	11.5	0/255	0.0	398/6,144	6.5
Lumbar	460/2,541	18.1	11/67	16.4	5/120	4.2	476/2,728	17.4
Total	1,026/11,017	9.3	53/447	11.9	12/536	2.2	1,091/12,000	9.1

[a]See Table 1 for references.

between the historic (9.8%) and precontact (9.6%) series (χ^2, $p > 0.05$). The frequencies of carious lesions are similar in the historic (12.0%) and precontact series (14.0%), a nonsignificant difference (χ^2, $p > 0.05$). The slightly higher dental caries frequencies in some precontact series may suggest a greater reliance on *poi*, a sticky, starchy staple that was traditionally consumed in great quantities. Although not reflected in these totals, most of the carious lesions are interproximal and neck caries rather than occlusal caries. Dental abscessing is significantly less in historic (3.4%) than in precontact (5.2%) Hawaiians (χ^2, $p \leq 0.05$).

Finally, calculus deposits, alveolar bone resorption, and severe attrition differ among the precontact and historic Hawaiian skeletons (Table 10). Calculus (tartar) found on the tooth at the gum-tooth interface acts as an irritant to the gums. Significantly higher levels of calculus are reported in the his-

TABLE 9. Comparison of Oral-Dental Pathological Conditions in Hawaiian Remains[a]

Skeletal series	Premortem tooth loss A/O	Premortem tooth loss %	Caries A/O	Caries %	Abscessing A/O	Abscessing %
Historic						
Kaka'ako, O'ahu	1/182	0.5	35/298	11.7	8/317	2.5
Auhaukea'e, Hawai'i	26/225	11.6	28/248	11.3	5/210	2.4
Wai'anae, O'ahu	24/116	20.7	13/85	15.3	8/90	8.9
Historic total	51/523	9.8	76/631	12.0	21/617	3.4
Historic/prehistoric						
'Anaeho'omalu, Hawai'i	92/1,165	7.9	19/437	4.3	49/890	5.5
Prehistoric						
Honokahua, Maui	852/8,897	9.6	985/7,309	13.5	381/7,629	5.0
Hā'ena, Kaua'i	61/571	10.7	92/443	20.8	15/381	3.9
Kualoa, O'ahu	29/455	6.4	58/397	14.6	29/385	7.5
Wai'anae, O'ahu	34/214	15.9	25/143	17.5	18/156	11.5
Prehistoric total	976/10,137	9.6	1,160/8,292	14.0	443/8,551	5.2

[a]See Table 1 for references.

toric (14.2%) series than in the precontact (6.5%) series (χ^2, $p \leq 0.01$). Alveolar resorption, the amount of tooth root exposed above the alveolar bone margin, is slightly, but not significantly greater, in the precontact (49%) than in the historic (42.8%) series (χ^2, $p > 0.01$). Thus, precontact Hawaiians had less calculus but more alveolar resorption than their historic counterparts. The prevalence of severe attrition—that is,

TABLE 10. A Comparison of Peridontal Disease and Attrition in Hawaiian Remains[a]

Skeletal series	Calculus[b] A/O	Calculus[b] %	Alveolar resorption[c] A/O	Alveolar resorption[c] %	Attrition[d] A/O	Attrition[d] %
Historic						
Kaka'ako, O'ahu	50/296	16.9	113/274	41.2	27/295	9.2
Auhaukea'e, Hawai'i	29/239	12.1	69/195	35.4	41/164	25.0
Wai'anae, O'ahu	9/85	10.6	60/97	61.9	13/89	14.6
Historic total	88/620	14.2	242/566	42.8	81/548	14.8
Historic/prehistoric						
'Anaeho'omalu, Hawai'i	84/445	18.9	217/443	49.0	125/438	28.5
Prehistoric						
Honokahua, Maui	489/7,167	6.8	3,721/7,194	51.7	3,456/7,365	46.9
Hā'ena, Kaua'i	23/427	5.4	28/297	9.4	109/442	24.7
Kualoa, O'ahu	13/428	3.0	84/303	27.7	128/438	29.2
Wai'anae, O'ahu	4/116	3.4	40/106	37.7	22/141	15.6
Prehistoric total	529/8,138	6.5	3,873/7,900	49.0	3,715/8,386	44.3

[a]See Table 1 for references.
[b]Includes calculus deposits scored as moderate and marked only.
[c]Includes expressions of alveolar resorption scored as moderate and marked only.
[d]Includes wear scored as reaching the dentin (moderate), pulp and root only.

dentin, pulp, and root exposure—is significantly greater in the precontact collections (44.3%) than in those dating to historic times (14.8%) (χ^2, $p \leqslant 0.01$).

Infectious Disease

Osseous responses to infection that vary greatly in severity have been reported for most precontact Hawaiian skeletal series, and they include osteolytic erosion, slight periostitis, osteitis, and septic arthritis. As is generally true of prehistoric skeletal series, most examples of precontact infectious disease are nonspecific in nature, although lesions consistent with tuberculosis and the treponematoses have been reported in some precontact series.

Coffin burials interred in Kaka'ako during the smallpox epidemic of 1853–1854 (Daws, 1968; Griffin et al., 1986) provide an unique picture of Hawaiian health. These individuals were from a population with little or no prior immunity to this acute, highly communicable, febrile, viral disease. The epidemic was started when the American ship *Charles Mallory* arrived in February 1853, with an infected sailor on board. By August, smallpox had reached epidemic proportions (Halford, 1954).

Smallpox itself does not affect the adult skeleton, but osteomyelitis and joint effusions can complicate the disease. In children, the occurrence of variolar osteomyelitis/arthritis varies from 2% to 20%, begins one to four weeks after the initial infection, typically involves the upper extremities (elbows), is often bilateral, and can damage the growth plate (Rang, 1969; Ortner and Putschar, 1981). At Kaka'ako, there is evidence of periosteal reactive bone in two individuals, an adult male and an adolescent. Differential diagnosis of the pathological changes seen in the long limb bones of the latter individual include smallpox and nonspecific infection (Pietrusewsky et al., 1989).

Another individual, a female aged 25–30 years old from Kaka'ako, has vertebral lesions consistent with tuberculosis. No signs of skeletal tuberculosis were noted in any of the remains from prehistoric Keōpū (Collins, 1986). Johnson and Kerley (1974:150) interpret vertebral lesions in two individuals from Mōkapu (Snow, 1974) as possible tuberculosis infections. Possible skeletal tuberculosis lesions have been described in 23 individuals from the Honokahua site (Pietrusewsky et al., 1991). These lesions, occurring in the putatively prehistoric Mōkapu and Honokahua collections, may reflect the arrival of new diseases associated with contact prior to noticeable changes in mortuary practices.

There are osseous changes in two or possibly three individuals from Kaka'ako suggestive of treponemal infections. The bone responses to these treponematoses (yaws and syphilis) are difficult, if not impossible, to distinguish from one another (Steinbock, 1976; Ortner and Putschar, 1981; Baker and Armelagos, 1988). There is only one published case of possible treponematosis from Hawai'i prior to Captain Cook's arrival, a young adult female from Barber's Point, O'ahu, dated to A.D. 1425–1650 (Pietrusewsky et al., 1990a). Incidentally, it is of interest to note that mitochondrial DNA extracted from this skeleton reveals the nine base-pair deletion seen frequently in Asians and Polynesians (Hagelberg and Clegg, 1993).

In the Honokahua series, there are 11 adults and four children with periostitic changes that may be associated with treponemal disease (Pietrusewsky et al., 1991), and six examples of subperiosteal new bone formation on lower shafts of tibiae were reported in the Mōkapu remains (Bowers, 1966). Where the lower leg is involved, however, it is difficult to eliminate other kinds of infections or injury in differential diagnoses.

DISCUSSION AND CONCLUSION

Skeletal data for precontact Hawai'i are reasonably comprehensive, although there will continue to be problems with determining the exact temporal provenience of skeletal remains, especially during the time immediately following first contact. The information for contact populations is less extensive but holds promise for assessing and interpreting the impact of Europeans on native Hawaiians. This chapter has shown significant differences between precontact and contact skeletal series for several health indicators. For many indicators, no differences through time could be detected. Interpreting such patterns will challenge students of Hawaiian skeletal biology, epidemiology, and paleopathology. Rigorous assessments of the demographic features of precontact and contact Hawai'i will prove elusive until larger, more representative, and better provenienced contact samples become available, and the basic assumptions behind life table construction are fully addressed by osteologists.

Our comparisons show little change in female stature from precontact Hawai'i to Hawaiians living ca. 1920. Precontact and postcontact Hawaiian males are about an inch taller than those measured in 1920. We emphasize that the differences are not great and may reflect, in part, the means of estimating stature from long bones and comparing the latter with measurements taken on living people.

A modest, but significant, increase in enamel hypoplasia frequencies was found in the contact compared to the precontact skeletal series. This increase is more likely due to an increased number of childhood insults than to an increased chance of survival in the historic series. Data on cribra orbitalia in contact Hawaiians are limited and, although the combined frequency is greater than that found in the precontact series, the differences between the two is not significant. Bone fractures are less common in contact than in precontact skeletal series, but here too the difference is not significant. Likewise, no significant differences between the early and late skeletal groups were found in degenerative changes in the articular surfaces and centra of vertebrae. The total frequency of osteoarthritis in the articular joint surfaces of the limb skeleton, however, was found to be significantly greater in the historic than in the precontact skeletal series.

Two indicators of dental pathology, dental caries and premorten tooth loss, reveal no significant differences over time. Calculus and alveolar resorption, however, were found to be significantly different in the two series. The most clear-cut difference in this survey is the more severe dental attrition in the precontact Hawaiians that suggests a coarser, less refined diet or an occupational use of teeth in prehistoric times.

Finally, the evidence of infectious diseases in Hawaiian remains is especially provocative and provides a glimpse of the effects of this monumental confrontation of cultures. The existence of treponemal disease (yaws or syphilis) in precontact Hawai'i has been a topic of continuing interest and debate (Pirie, 1978; Stannard, 1989). Treponemal infection, undoubtedly yaws, has been demonstrated in precontact populations in western Polynesia (Pietrusewsky, 1969) and in the Mariana Islands (Stewart and Spoehr, 1952; Pietrusewsky and Douglas, 1989c). In Hawai'i, there is only one example of a treponemal infection (Pietrusewsky et al., 1990a) from a clearly precontact situation, and the lesions in this specimen are especially distinctive. The other examples of such infections, as well as skeletal tuberculosis, come from Honokahua and Mōkapu, both prehistoric in provenience. It is important to note, however, that "prehistoric" is a relative term and both sites are likely to have been used in the premissionary phase (1778–1820) of the early contact era.

Although the skeletal evidence for assessing the health of Hawaiian populations during the early contact period is scarce and skeletal series are typically small and represent short temporal sequences, this initial investigation suggests that there are indications of a deterioration in health. The extreme impact of some venereal diseases (such as gonorrhea) introduced at first Western contact on the fertility of the Hawaiian people are not readily identifiable in the skeletal record. Other infectious diseases, however, such as tuberculosis and treponematosis (yaws and syphilis), traditionally assumed to have been introduced to Hawai'i with Western contact, are now either confirmed or challenged by the osteological record.

ACKNOWLEDGMENTS

First, our sincere thanks to the editors of this volume who invited us to contribute a chapter on Hawai'i and whose constant support and encouragement made this task a reality. Because much of the information presented in this chapter derives from work in contract archaeology, we would further like to acknowledge the following cultural resource management firms in the State of Hawai'i who have assisted in building this database: Cultural Surveys Hawai'i, International Archaeological Research Institute, Inc., Ogden Environmental and Energy Services, Co., Inc., and Paul H. Rosendahl, Ph.D., Inc. Our further thanks to Rona Michi Ikehara and Darren Miwa for their assistance in manuscript preparation and to Mark Nakamura for graphic illustration. And finally, it is with deepest respect, that we thank the *kupuna* of long ago, who have permitted us to learn so much about them from their *'iwi*.

REFERENCES CITED

Allen H (1898) A study of Hawaiian skulls. Wagner Free Institute of Science 5:1–55.

Baker B, and Armelagos G (1988) The origin and antiquity of syphilis: Paleopathological diagnosis and interpretation. Current Anthropology 29:703–737.

Beaglehole JC (1967) The Journals of Captain James Cook on His Voyage of Discovery, vol. 3, parts 1 and 2. Cambridge: Cambridge University Press.

Bocquet-Appel JP, and Masset C (1982) Farewell to paleodemography. Journal of Human Evolution 11:321–333.

Bowers WF (1966) Pathological and functional changes found in 864 pre-Captain Cook contact Polynesian burials from the sand dunes of Mōkapu, Oahu, Hawaii. Internal Surgery 45: 206–217.

Carter LA (1990) Protohistoric Material Correlates in Hawaiian Archaeology. M.A. thesis, Department of Anthropology, University of Hawai'i, Mānoa.

Collins SL (1986) Osteological studies of human skeletal remains from the Keōpū burial site. In TL Han, SL Collins, DS Clark, and A Garland (eds.): *Moe Kau a Ho'oilo*: Hawaiian Mortuary Practices at Keōpū, Kona, Hawai'i. Departmental Report Series. Report 86-1. Honolulu: Department of Anthropology, B.P. Bishop Museum, pp. 165–249.

Davis JB (1867) Thesaurus Craniorum: Catalogue of the Skulls of the Various Races of Man in the Collection of Joseph B. Davis, London.

Daws G (1968) Shoal of Time: A History of the Hawaiian Islands. New York: Macmillan.

Goodman BM, Armelagos GJ, and Rose J (1980) Enamel hypoplasias as indicators of stress in three prehistoric populations from Illinois. Human Biology 52:515–528.

Griffin PB, Keene DTP, and Kennedy J (1986) Kaka'ako: Prediction of Subsurface Archaeological Resources. Kaka'ako Community Development District-Archaeological Reconnaissance Survey. Unpublished manuscript.

Hagelberg E, and Clegg JB (1993) Genetic polymorphisms in prehistoric Pacific Islanders determined by analysis of ancient bone DNA. Proceedings of the Royal Society of London B 252: 163–170.

Halford FJ (1954) Nine Doctors and God. Honolulu: University of Hawai'i Press.

Houghton P, Leach BF, and Sutton DG (1975) The estimation of stature of prehistoric Polynesians in New Zealand. Journal of the Polynesian Society 84:325–336.

Howell N (1982) Village composition implied by a paleodemographic life table: The Libben site. American Journal of Physical Anthropology 59:263–269.

Johnson LC, and Kerley ER (1974) Report on pathological specimens from Mōkapu. In CE Snow (ed.): Early Hawaiians: An Initial Study of Skeletal Remains from Mōkapu, Oahu. Lexington: The University Press of Kentucky. Appendix B, pp. 149–158.

Kam W (1979) Osteological study of human skeletal remains at Kālahuipua'a. In PV Kirch: Marine Exploration in Prehistoric Hawai'i. Pacific Anthropological Records 29, pp. 211–228.

Kamakau SM (1964) Ka Po'e Kahiko: The People of Old. B.P. Bishop Museum Special Publication 51. Honolulu: Bishop Museum Press.

Kirch PV (1984) Evolution of the Polynesian Chiefdoms. Cambridge: Cambridge University Press.

Kirch PV (1985) Feathered Gods and Fishhooks: An Introduction to Hawaiian Archaeology and Prehistory. Honolulu: University of Hawai'i Press.

Lukacs JR (1989) Dental paleopathology: Methods for reconstructing dietary patterns. In MY Işcan and KA Kennedy (eds.): Reconstruction of Life from the Skeleton. New York: Alan R. Liss, pp. 261–286.

Ortner DJ, and Putschar WGJ (1981) Identification of Pathological Conditions in Human Skeletal Remains. Smithsonian Contributions to Anthropology 28.

Pietrusewsky M (1969) An osteological study of cranial and infracranial remains from Tonga. Records of the Auckland Institute and Museum 6:287–402.

Pietrusewsky M (1971) Human Skeletal Remains at 'Anaeho'omalu. Departmental Report Series 7-17. Honolulu: Department of Anthropology, B.P. Bishop Museum.

Pietrusewsky M, and Douglas MT (1989a) An Osteological Study of Human Remains Recovered from Kualoa Regional Park, Windward O'ahu Island, Hawai'i. Unpublished manuscript.

Pietrusewsky M, and Douglas MT (1989b) An Osteological Study of Late Historic Human Remains from 'Auhaukea'ē, North Kona, Hawai'i. Unpublished manuscript.

Pietrusewsky M, and Douglas MT (1989c) Human Remains from Oleai, Saipan, Commonwealth of the Northern Mariana Islands. Unpublished manuscript.

Pietrusewsky M, and Douglas MT (1990) An Osteological Study of Human Remains Recovered in 1988–89 from Wai'anae Army Recreation Center, Wai'anae, O'ahu, Hawai'i. Unpublished manuscript.

Pietrusewsky M, and Douglas MT (1992) Tooth evulsion in ancient Hawai'i. American Journal of Physical Anthropology Supplement 14:131–132.

Pietrusewsky M, Douglas MT, and Ikehara RM (1989) An Osteological Study of Human Remains Recovered from South Street and Quinn Lane, Kaka'ako, O'ahu, Hawai'i. Unpublished manuscript.

Pietrusewsky M, Douglas MT, and Ikehara RM (1990a) A human skeleton found at Barber's Point, Hono'uli'uli, 'Ewa, O'ahu. In B Davis: Archaeological and Palaeontological Investigations at the Proposed HECO Barber's Point Generating Station, Hono'uli'uli, 'Ewa, O'ahu. Honolulu: International Archaeological Research Institute, Inc. Report, Appendix C, pp. 92–109.

Pietrusewsky M, Douglas MT, and Kalima PA (1990b) Human Skeletal Remains Recovered from 'Anaeho'omalu, South Kohala, Hawai'i Island: A Second Study. Paul H. Rosendahl, Ph.D. Inc. Report No. 545-091990.

Pietrusewsky M, Douglas MT, Kalima PA, and Ikehara RM (1991) Human Skeletal and Dental Remains from the Honokahua Burial Site, Land of Honokahua, Lahaina District of Maui, Hawai'i. Paul H. Rosendahl, Ph.D. Inc. Report No. 246-041091.

Pietrusewsky M, Douglas MT, and Ikehara RM (1992) Osteological Analysis of Human Remains Found at Site (50-30-03-1809) (TMK 5-9-02-31) Hā'ena, Halele'a, Kaua'i (Anawalt Property). Unpublished manuscript.

Pirie M (1978) The Consequences of Cook's Hawaiian Contacts on the Local Population. In JN Hurd and M Kodama (eds.): Captain Cook and the Pacific Islands. University of Hawaii Pacific Islands Program. Miscellaneous Working Papers 3, pp. 75–88.

Rang M (ed.) (1969) The Growth Plate and its Disorders. London: E. & S. Livingstone, Ltd.

Robbins, S (1967) Pathology, 3rd ed. Philadelphia: WB Saunders.

Sattenspiel L, and Harpending H (1983) Stable populations and skeletal age. American Antiquity 48:489–498.

Snow CE (1974) Early Hawaiians: An Initial Study of Skeletal Remains from Mōkapu, O'ahu. Lexington: University of Kentucky Press.

Stannard DE (1989) Before the Horror: The Population of Hawai'i on the Eve of Western Contact. Honolulu: Social Sciences Research Institute, University of Hawai'i.

Steinbock RT (1976) Paleopathological Diagnosis and Interpretation: Bone Diseases in Ancient Human Populations. Springfield: Charles C Thomas.

Stewart TD, and Spoehr A (1952) Evidence on the paleopathology of yaws. Bulletin of the History of Medicine *26*:538–553.

Stuart-Macadam P (1987) Porotic hyperostosis: New evidence to support the anemia theory. American Journal of Physical Anthropology *74*:521–526.

Sullivan LR (1927) Observations on Hawaiian Somatology. Memoirs of B.P. Bishop Museum IX, No. 4.

Suzuki T (1987) Cribra orbitalia in the early Hawaiians and Mariana Islanders. Man and Culture in Oceania *3*:95–104.

Thomas DH (1986) Refiguring Anthropology: First Principles of Probability and Statistics. Prospect Heights: Waveland Press.

Uhde CWF (1861) Über die Schädelform der Sandwich-Insulaner Verhandlungen der Kaisolichen Leopoldino-Carolinischen Deutschen. Akademie der Naturforscher *28*:3–12.

Underwood JH (1969) Human Skeletal Remains from Sand Dune Site (H1), South Point (Ka Lae), Hawaii. Pacific Anthropological Records 9.

Walker PL (1989) Cranial injuries as evidence of violence in prehistoric southern California Indians. American Journal of Physical Anthropology *80*:313–323.

Weiss KM (1973) Demographic Models for Anthropology. Memoirs of the Society of American Anthropology 27.

Index